T0338490

MODERN LENS ANTENNAS FOR COMMUNICATIONS ENGINEERING

MODERN LENS ANTENNAS FOR COMMUNICATIONS ENGINEERING

John Thornton

Kao-Cheng Huang

IEEE PRESS

WILEY

Cover Design: John Wiley & Sons, Inc.
Cover Illustration: © simon2579/iStockphoto.

Published by John Wiley & Sons, Inc., Hoboken, New Jersey.
Published simultaneously in Canada.

For general information on our other products and services or for technical support, please contact our Customer Care Department within the United States at (800) 762-2974, outside the United States at (317) 572-3993 or fax (317) 572-4002.

Wiley also publishes its books in a variety of electronic formats. Some content that appears in print may not be available in electronic formats. For more information about Wiley products, visit our web site at www.wiley.com.

Library of Congress Cataloging-in-Publication Data is available.

ISBN: 978-1-118-01065-5

Printed in the United States of America.

10 9 8 7 6 5 4 3 2 1

CONTENTS

PREFACE

The aim of this book is to present the modern design principles and analyses of lens antennas. It gives graduates and RF/microwave professionals the design insights in order to make full use of lens antennas. The reader might ask: Why is such a book considered necessary and timely? The reply we would bring to such an inquiry is that the topic has not been thoroughly publicized recently and so its importance has become somewhat underestimated. Furthermore, the work has brought about an opportunity to gather together the authors' contributions to several areas of research where lens antennas have been promoted. Foremost among these are communications applications, where of course antennas play a key role and where we will show why certain advantages accrue from the particular characteristics of lens antennas.

The major advantages of lens antennas are narrow beamwidth, high gain, low sidelobes and low noise temperature. Their structures can be more compact and weigh less than horn antennas and parabolic reflector antennas. Lens antennas, with their quasi-optical characteristics, also have low loss, particularly at near millimeter and submillimeter wavelengths where they have particular advantages. Beam shaping can be achieved by controlling the phase distribution across the lens aperture in a manner that can be more accurate and less costly than would be the case for a reflector. Such a shaped dielectric lens can be more economical to produce in small- to medium-scale production runs than other antenna types where certain niche applications are considered. In addition, spherical lens antennas have the benefit of no scan loss and wide bandwidth, with the option for multiple beams from a common aperture.

Modern Lens Antennas for Communications Engineering serves as an excellent tool for RF/microwave professionals (engineers, designers, and developers) and industries with microwave and millimeter wave research projects. For university students, this book requires a prerequisite course on antennas and electromagnetic waves, which covers propagation, reflection, and transmission of waves, waveguides, transmission lines, and some other antenna fundamental concepts. Such a course is usually followed by design projects. This book can be used as further study material in such design projects. Advanced students and researchers working in the field of modern communications will also find this book of interest. Included is a bibliography of current research literature and patents in this area.

Based on these credentials, this book systematically conducts advanced and up-to-date treatment of lens antennas. It does not purport to present a far-reaching treatise on every aspect of lens antennas, but rather, following the introductory chapters, the emphasis of the work is taken from the authors' own research of recent years. Example designs are presented and the analysis of their performance detailed.

A summary of each chapter is as follows.

Chapter 1 gives an overview of different types of lens antennas and their history. It discusses basic principles of delay lenses (in which the electrical path length is increased by the lens medium), fast lenses (in which the electrical path length is decreased by the lens medium), materials for lenses, and applications for lens antennas. It attempts a fairly broad review of some quite disparate antenna types that are nevertheless classed as "lenses" such as the planar or frequency selective surface type and also the Fresnel zone variants of dielectric lenses. Antenna measurement techniques are also summarized.

Chapter 2 reviews important wave propagations and antenna parameters, for the purpose of consistency in notation and easy referencing. The material progresses from uniform plane waves in various media, such as lossless and lossy dielectrics, to all important antenna parameters.

Chapter 3 focuses on low-cost yet high-directivity dielectric polyrod antennas. Different feeding methods, maximum gain, and beam tilting are discussed in detail. A multibeam polyrod array is presented where this increases radiation coverage, and phase compensation is introduced to adjust beam direction.

Chapter 4 tackles millimeter wave issues such as high path loss and high power consumption. It then explores the variety of millimeter wave lens antennas and novel design methods. Quasi-optical characteristics of lens antennas are identified for aiding design at millimeter and submillimeter wavelengths.

Chapter 5 discusses the properties of antennas which would be required for communications from high-altitude platforms. As such, it presents a case study where lens antennas were identified as being a potential solution for this niche area. Beginning with a system-level analysis of a cellular architecture employing spectrum reuse through multiple spot beams, the chapter goes on to show how a type of lens antenna, with shaped beam and low sidelobes, directly controls cochannel interference. A practical design and results are reported.

Chapter 6 presents a summary of the properties of spherical lens antennas including the Luneburg lens and its relatives. Analytical techniques are discussed, beginning with ray tracing but then leading to the much more powerful spherical wave expansion technique. Lens construction problems are addressed, and then the properties of constant-index spherical lenses are summarized.

Chapter 7 follows on from Chapter 6 and reports from several programs where hemispherical lens-reflector antennas were developed in practice. Here, a hemisphere with ground plane recovers the equivalent aperture of a spherical lens but in half the height—a profound advantage where a low profile is required. A dual-beam lens antenna for satellite communications is reported, as is a constant-index lens reflector partially developed for aircraft-to-ground links.

JOHN THORNTON
KAO-CHENG HUANG

ACKNOWLEDGMENTS

The two authors contributed equally to this book. During the period of the manuscript's preparation, the authors have been obliged to many people. First of all, the authors wish to acknowledge the valuable comments from the reviewers. Also, the authors would like to acknowledge the copyright permission from IEEE.

The authors also wish to thank Dr. Derek Gray for his contribution to Section 7.7.

The authors are indebted to many researchers for their published works, which were rich sources of reference. Their sincere gratitude extends to IEEE editors for their support in writing the book. The help provided by Taisuke Soda, Mary Hatcher, and other members of the staff at John Wiley & Sons is most appreciated.

In addition, Kao-Cheng Huang would like to thank Prof. David J. Edwards, University of Oxford (United Kingdom), Prof. Zhaocheng Wang, Tsinghua University (China), Prof. Mook-Seng Leong, National University of Singapore (Singapore), Prof. Zhichun Lei, University of Applied Sciences Ruhr West (Germany), Prof. Hsueh-Man Shen, New York University (United States), Prof. Jia-Sheng Hong, Heriot-Watt University (United Kingdom), and Miss Hsiang-Jung Huang, Illinois Institute of Technology (United States) for their many years of support. Dr. Huang would also like to acknowledge the copyright permission from John Wiley & Sons, Inc. (Chapters 5 and 6 of *Millimetre Wave Antennas for Gigabit Wireless Communications: A Practical Guide to Design and Analysis in a System Context* by Huang and Edwards, 2008), Springer (Chapter 17 in *Antenna Handbook*, Vol. 3: *Antenna Applications* by Lo and Lee, 1993), and Sophocles J. Orfanidis (Sections 1.5 and 2.12 in *Electromagnetic Waves and Antennas* by Sophocles J. Orfanidis, 2010), for extracts used in Chapters 2, 3, and 4.

John Thornton thanks the following for supporting the work described in Chapters 5–7.

At the University of York, Dr. David Grace (project manager, EU FP6 CAPANINA Project 2003–2006) and Mr. Tim Tozer (leader of Communications Research Group).

The middle part of Chapter 7 describes the European Space Agency-funded Multiscan project under Contract Number 20836/07/NL/CB during 2007–2009. The Agency's support is gratefully acknowledged as is that of program manager Maarten van der Vorst.

Thanks to Andy White and Andy Patterson for bringing that project to fruition, to Mark Hough for measurements and logistics support, and to Graham Long for mechanical design. Thanks to Philip Haines of Hollycroft Associates for contributing an industrial perspective. An article describing that project is also found in the August 2009 edition of *Microwave Journal*.

Dr. Thornton also thanks Derek Gray for coauthoring several papers on lens theory in the late 2000s and Stuart Gregson (Nearfield Systems Inc., Torrance, California) both for permission to use the scanner image in Chapter 1 and for moral support with authorship.

J.T.
K.-C.H.

1

INTRODUCTION

John Thornton and Kao-Cheng Huang

The topic of lens antennas was widely investigated during the early development of microwave antennas and was influenced by the extensive body of existing work from optics. Subsequently, interest declined somewhat as lens antennas were overtaken by reflectors for high efficiency, large aperture antennas; and by arrays for shaped-beam, multi-beam, and scanning antennas. Quite recently, as research interest has expanded into the use of millimeter wave and sub-millimeter wave frequency bands, lens antennas have again attracted developers' attention.

This chapter is organized as nine sections to introduce the basics of lens antennas. Section 1.1 gives an overview of lens antennas, including its advantages, disadvantages, and the materials encountered. This is followed by a discussion of antenna feeds at Section 1.2. Then Section 1.3 introduces the fundamentals of the Luneburg lens (a topic to which Chapters 6 and 7 are dedicated). Section 1.4 introduces quasi-optics and Section 1.5 treats design rules. A discussion of metamaterials for lens antennas makes up Section 1.6 and then the planar lens array, which is a relative of the reflect-array antenna, follows in Section 1.7. Applications are proposed in Section 1.8 and measurement techniques and anechoic chambers discussed in the final section.

Modern Lens Antennas for Communications Engineering, First Edition. John Thornton and Kao-Cheng Huang.
© 2013 Institute of Electrical and Electronics Engineers. Published 2013 by John Wiley & Sons, Inc.

1.1   LENS ANTENNAS: AN OVERVIEW

The use of dielectric lenses in microwave applications seems to date back to the early days of experiments associated with the verification of the optical properties of electromagnetic waves at 60 GHz [1]. However, it was not until World War II that lenses gained interest as antenna elements. Even then they were not widely used because of their bulky size at rather low frequencies.

Nowadays there is a renewed interest in dielectric lenses, not least because of the rapidly growing number of applications for millimeter waves where lens physical dimensions have acceptable sizes. Besides, very low loss dielectric materials are available, and present-day numerically controlled machines enable low-cost fabrication of quite sophisticated lenses made with very good tolerances.

In one of the earliest dielectric lens antenna applications, a homogeneous lens was designed to produce a wide-angle scanning lobe [2]. Also, homogeneous lenses have been used as phase front correctors for horns. The lens is often mounted as a cap on a hollow metallic horn [3]. In this configuration the lens surfaces on both sides can be used to design for two simultaneous conditions. In addition, lenses may be designed to further control the taper of the field distribution at the lens aperture [4] or to shape the amplitude of the output beam in special applications [5].

The aperture of a solid dielectric horn can be shaped into a lens to modify or improve some radiation characteristics [6]. For instance, the aperture efficiency of a solid dielectric horn may be improved by correcting the aperture phase error. Alternatively we may use a lens to shape the amplitude of the output beam or to improve the cross-polarization performance, but because there is only the one lens surface to be varied, only one of these design targets might be made optimum.

1.1.1   The Microwave Lens

In optics, a lens refracts light while a mirror reflects light. Concave mirrors cause light to reflect and create a focal point. In contrast, lenses work the opposite way: convex lenses focus the light by refraction. When light hits a convex lens, this results in focusing since the light is all refracted toward a line running through the center of the lens (i.e., the optical axis). Save for this difference, convex lens antennas work in an analogous fashion to concave reflector antennas. All rays between wavefronts (or phase fronts) have equal optical path lengths when traveling through a lens. Fresnel's equations, which are based on Snell's law with some additional polarization effects, can be applied to the lens surfaces.

In general, lenses collimate incident divergent energy to prevent it from spreading in undesired directions. On the other hand, lenses collimate a spherical or cylindrical wavefront produced respectively by a point or line source feed into an outgoing planar or linear wavefront. In practice, however, complex feeds or a multiplicity of feeds can be accommodated since performance does not deteriorate too rapidly with small off-axis feed displacement.

There are two main design concepts used to reach different goals.

1. Conventional (e.g., hyperbolical, bi-hyperbolical, elliptical, hemispherical) or shaped lens antennas are used simply for collimating the energy radiated from a feed.
2. In the case of shaped designs, more complex surfaces are chosen for shaping the beam to produce a required radiation pattern, or for cylindrical and spherical lenses for beam scanning with either single or multiple feeds.

Lenses can also be placed into one or other of the categories of slow wave and fast wave lenses (Fig. 1.1). The terms relate to the phase velocity in the lens medium. The slow-wave lens type is illustrated in Figure 1.1a. Here, the electrical path length is increased by the medium of the lens, hence the wave is retarded.

The most common type is the dielectric lens but another example is the *H*-plane metal-plate lens (Fig. 1.2). (The *H*-plane is that containing the magnetic field vector

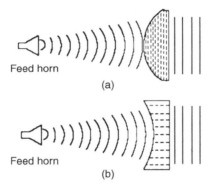

Figure 1.1. Comparison of (a) slow-wave or dielectric lens and (b) *E*-plane metal-plate (fast-wave) lens types. Wave fronts are delayed by (a) but advanced by (b).

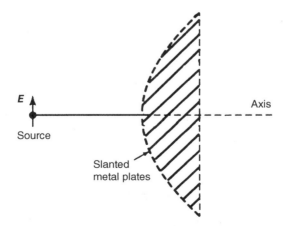

Figure 1.2. *H*-plane metal plate lens antenna.

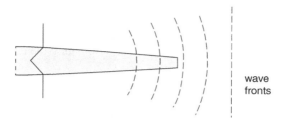

Figure 1.3. A form of basic lens antenna: the polyrod.

and the also the direction of maximum radiation, or main lobe boresight. The magnetizing field or *H*-plane lies at a right angle to the electric or *E*-plane.)

Figure 1.1b shows the fast-wave lens type where the electrical path length is effectively made shorter by the lens medium, so the wave is advanced. *E*-plane metal-plate lenses are of the fast-wave type. (The *E*-plane is that containing the electric field vector and also the direction of maximum radiation. The *E*-plane and *H*-plane are orthogonal to each other and determine the polarization sense of a radio wave.)

In terms of materials, dielectric lenses may be divided into two distinct types:

1. Lenses made of conventional dielectrics, such as Lucite® or polystyrene.
2. Lenses made of artificial dielectrics such as those loaded by ceramic or metallic particles.

Lens antennas tend to be directive rather than omnidirectional i.e. they usually exhibit a single, distinct radiation lobe in one direction. In this case, they might be thought of as "end-fire" radiators. With this in mind, a dielectric rod, as shown in Figure 1.3, is a good example. Because the rod is usually made from polystyrene, it is called a *polyrod*. A polyrod acts as a kind of imperfect or leaky waveguide for electromagnetic waves. As energy leaks from the surface of the rod it is manifested as radiation. This tendency to radiate is deliberate, and the rod's dimensions and shape are tailored to control the radiation properties which are discussed in proper detail in Chapter 3.

In contrast to polyrod antennas, most lens dimensions are much larger than the wavelength, and design is based on the quasi optics (QO) computations. Snell's refraction law and a path length condition (or eventually an energy conservation condition) are then used to define the lens surface in the limit as the wavelength tends to zero. Depending on the lens shape, diffraction effects may give rise to discrepancies in the final pattern. While these comments really concern axis-symmetric lenses, on the other hand, literature on arbitrary shaped dielectric lenses and three-dimensional amplitude shaping dielectric lens is also available [5].

1.1.2 Advantages of Lens Antennas

Lenses are an effective antenna solution where beam shaping, sidelobe suppression, and beam agility (or steering in space) can be achieved simultaneously from a compact assembly. Dielectric lenses may also bring about these system advantages economically because they can be manufactured through molding or automated machining in reasonably high quantities, and these processes offer suitable tolerances. Lenses are used to convert spherical phase fronts into planar phase fronts across an aperture to enhance its directivity much like parabolic reflectors. For this purpose lenses present an advantage over reflectors in that the feed is located behind the lens, thus eliminating aperture blockage by the feed and supporting struts, with no need for offset solutions. This ability to illuminate the secondary aperture (lens vs. reflector) in a highly symmetric fashion but without aperture blockage leads to the benefit that lens antennas can have low distortions and cross-polarization.

Microwave lenses are of course not without disadvantages and among these are dielectric losses and reflection mismatch. A mitigation against loss of course is to use low loss materials, such as Teflon, polyethylene, and quartz, where material loss can be reduced to quite negligible levels. Solutions for reflection mismatch include the use a quarter wavelength anti reflection (AR) layer, or "coating" if such a layer can be said to be thin. To follow up here with some examples, alumina-loaded epoxy has been used with good results as an anti-reflection coating for silicon lenses [7]. However, the epoxy material suffers from large absorption loss above 1 THz [8]. Englert et al. [9] has successfully coated both sides of a silicon window with 20 μm of low-density polyethylene (LDPE, $n \sim 1.52$) to achieve anti-reflection (AR) performance at $\lambda = 118$ μm. Other common plastics such as *Mylar* and *Kapton* are potential candidates because their refractive indices are close to the required value of $(n_{silicon})^{0.5} \sim 1.85$ [10]; however, such materials may be difficult to apply to a small silicon lens. Thin films of parylene can be used as an AR layer for silicon optics and show low-loss behavior well above 1 THz [10]. In contrast, at lower frequencies, for example, 10 to 60 GHz, air grooves may be machined into a lens surface to yield an effective air-dielectric composite layer of intermediate index—such a fabrication technique is encountered in commercially available lens antennas. This is also said to reduce gain variation with frequency.

1.1.3 Materials for Lenses

Real-life antennas make use of real conductors and dielectrics. It is thus useful to recall some of their characteristics. A conductor is defined as a material with a large number of free detachable electrons or a material having high conductivity. Typical conductors used in antennas are

- Silver (conductivity = 6.14×10^7/ohm/m)
- Copper (conductivity = 5.8×10^7/ohm/m)
- Aluminum (conductivity = 3.54×10^7/ohm/m)

A dielectric is essentially an insulator: a material with few free detachable electrons or with a low value of conductivity. In many cases conductivity for a good conductor may be taken as infinity and for a good dielectric as zero.

To produce lens antennas, it is necessary to select a material which is mechanically and electromagnetically stable. A typical choice of material could be based on a relative permittivity ranging from about 1.2 to 13—indeed these materials are either in use today or are expected to be used for millimeter wave antenna systems. Because of time and space considerations, not all materials available today will be compared here, but it is felt that these examples are representative and other common substrate materials have properties roughly within the range of those considered here.

The dielectric loss factor is also known as the dissipation factor. It is defined from the tangent of the loss angle $(tan\,\delta)$, and is hence also called the loss tangent. Put another way, the loss angle is that whose tangent is derived from the ratio of the imaginary to the real part of the dielectric permittivity: Im (ε)/Re (ε).

Thus we are quantifying the ratio of the resistive (also lossy, and imaginary) and the reactive (also lossless, and real) components of the dielectric constant. For low loss materials the former term should be very small. The dielectric constant (ε) and loss tangent $(tan\,\delta)$ can be difficult to measure accurately at microwave frequencies. Often, techniques are used which look at the properties of resonators and might compare their behavior with and without the insertion of a sample of dielectric material. Vector network analyzer would typically be used. Some typical values are identified in Table 1.1.

1.1.4   Synthesis

Synthesis of lenses has received plenty of attention, though perhaps less so than reflector antennas or other types. Spherical lenses (Fig. 1.4) present a good example of how synthesis and optimization can be applied. The topic will be covered in more detail in Chapters 6 and 7, but for present section it is worth pointing out that the ideal spherical lens configuration—the Luneburg lens—is exceedingly difficult to construct according to the strict formulation. When approximations are applied, by using discrete dielectric layers for example, compromises must be sought. These trade the antenna efficiency against the number of layers and hence difficulty of construction [12]. Modern computers assist greatly with the study of these effects. Computer optimization routines are widely applied in electromagnetics and microwave engineering. Different niche approaches have advantages and disadvantages depending on their precise area of application. In Reference 12 genetic algorithms were reported to have had powerful effect in optimizing the desired lens gain and sidelobe suppression, where the optimizer variables were the dielectric constant and radial width of each lens layer.

Komljenovic et al. [13] pointed out that of the many global optimization techniques available, those used most extensively so far in electromagnetics have been relatively few. These have been:

- genetic algorithms,
- particle swarm optimization.
- the multidimensional conjugate gradient method.

The genetic algorithm (GA), as a class of optimization kernel, has been applied widely and with success to many problems [14] in addition to the spherical lens we have already briefly mentioned. The GA approach is also know of as evolutionary computation, or is sometimes synonymous with the general class of evolutionary algorithms

TABLE 1.1. Electrical Properties of Microwave Substrates [11]

Loss tangent

Permittivity	0.0001	0.0002	0.0003	0.0004	0.0005	0.0006	0.0007	0.0008	0.0009	0.001	0.002	0.005	0.010	0.050	0.100
1															
1.2														$MgCO_3$	
2															
2.1	Teflon														
2.2				Polypropylene											
2.3		Polyethylene													
2.4															
2.5		Polystyrene													
3		Quarz													
4			BN												
5			Mica												
6				BeO											
7											GreenTape943				
8															
9			Sapphire MgO			Al_2O_3									
10													$MgTiO_3$		
12.5										GaAs					

7

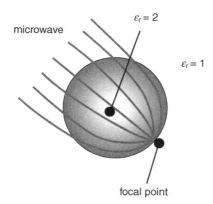

microwave

$\varepsilon_r = 2$

$\varepsilon_r = 1$

focal point

Figure 1.4. An example of a spherical stepped-index lens.

inspired by biological processes. In this context, we encounter such mechanisms as selection, mutation and inheritance, and the health of populations and so on.

A more recent technique, also under the "evolutionary algorithm" banner, is called "particle swarm" optimization [15]. This is based on models for the intelligence and movement of individuals in swarms and has been applied successfully to problems in electromagnetics. At first, relatively small numbers of individual states, perhaps fewer than 30, were considered adequate to model swarms. More recently though, and for more exhaustive optimization, larger numbers than this have been recommended, for example, in Reference 13, where the technique was applied to cylindrical lenses. Unsurprisingly, the computational load should be expected to increase as optimization techniques become ever more complex. Fortunately, computer performance is ever advancing.

On a final note for this section, a multidimensional conjugate gradient method has also been used for the successful optimization of some quite complicated shaped lens designs which have also had arbitrary geometries [16]. Again, the technique is iterative.

1.2   FEEDS FOR LENS ANTENNAS

The feeding methods of lens antennas can be any other type of antenna: horns, dipoles, microstrip (e.g., patch), and even arrays of antenna elements.

In practice horns (or just open ended waveguides) and patches are most commonly used, or, in some cases, arrays of such elements.

1.2.1   Microstrip Feeds

The most often cited advantages of microstrip (or "patch") antennas include their low mass, ease and low cost of fabrication, robustness, and ease of integration with other microwave printed circuits and connections. They are typically also low profile, occupying either just the thickness of the printed circuit, or the laminate plus a low dielectric constant spacing layer, which might just be air. The latter approach tends to increase the bandwidth of what is otherwise a narrow-band radiator [17].

As a primary feed for a lens antenna, the properties of patch antennas are useful in some cases but not in others. For example, where a patch antenna is used to illuminate a hemispherical lens-reflector, which is a type of reduced height scanning antenna, the low profile of the patch can help to avoid the headroom encroachment that would be caused by a waveguide or horn type feed. However, disadvantages of the patch include ohmic loss in both the metallization layer and the substrate, which can increase dramatically with frequency.

Many single-patch geometries can be used, for example, rectangular, circular, bow-tie, planar-conical, and so on. Where these can be combined in array configurations there is more scope to select or optimize the pattern which illuminates the so-called secondary aperture (the lens).

In Chapter 4 we will show some design examples of types of lens integrated directly onto patch antenna substrates.

1.2.2 Horn Feeds

In contrast to the narrow-band and low-efficiency printed circuit antenna discussed above, the waveguide or horn type tends to be both much more wide-band and more efficient. These advantages tend to accrue at the cost of physical volume, mass, and to some extent fabrication expense compared to patches at least. Considering that dielectric lenses (i.e., not zoned for reduced thickness) are very wide band in operation it can make sense to use them with a similarly wide-band primary feed. The horn's bandwidth is, to a first order at least, determined by that of the feeding waveguide and this will exhibit a cut-off behavior at the lower threshold and higher order mode propagation at a higher threshold. Used as a primary feed for a lens, the horn is carrying out a very similar function as when used with a reflector, but with the lens the horn will not introduce aperture blockage. In Chapter 5 we will present a low sidelobe lens antenna where this property is exploited.

Due to the lower propagation velocity, phase errors at the plane aperture of a solid dielectric horn are higher than for a conventional metallic horn of comparable size, thus placing a lower limit to the maximum achievable gain. Following a standard approach for high-performance metallic horns, where a dielectric lens is positioned at the horn aperture in order to correct the phase error, the dielectric horn aperture may be shaped into a lens [18].

Here, a practical engineering tip is that the feed should provide an illumination of the antenna edge at the level of $-10\,dB$ with respect to the central point. Then, the antenna performance in terms of directive gain is optimal. This empirical rule is approximately valid for any QO system, although for systems carrying out extensive manipulations of Gaussian beams a more conservative rule of $-20\,dB$ to $-35\,dB$ is common [19].

There are other applications where the design is not necessarily to maximize antenna gain, but rather to shape the beam. Again, a lens may be used to modify the basic horn radiation pattern. A particular design goal is discussed in detail in Chapter 5 where the case study is for a low sidelobe elliptic beam antenna for mm-wave communications.

Intrinsic to the dielectric horn-lens geometry, there is only one refracting surface and so amplitude shaping and phase correction conditions can not be imposed simultaneously. The situation is different for metallic horns where correction lenses have two refracting surfaces, one on the horn side and the other on the output (free space) side, enabling two independent conditions to be met. In this case, however, rather than designing the second refracting surface for an amplitude condition it is often better to set the input surface as planar and to design the output surface according to the path length condition, since an appropriate choice of permittivity leads to convenient field distribution on the aperture [18].

Another design approach can be to employ moderate loss materials to reduce the field amplitude toward the lens axis where its depth increases, thus reducing the field taper across the aperture without interfering with the other lens surface [20] at the expense of gain.

Horn-lenses are just one possible application of dielectric lenses. Most commonly used are axis-symmetric lenses, designed as collimating devices possibly with more than one focal point for scanning and multibeam applications. In the simplest designs for single focus lenses one of the two lens surfaces is arbitrarily fixed to some preferred shape while phase correction condition defines the other surface. For scanning and multibeam applications the second surface is used to introduce a further condition to minimize aberrations originated by off-axis displacements of the feed position.

The main shortcoming of horns lies in their bulkiness that can make integration challenging. A less bulky alternative would be an open ended waveguide, but this would reduce optimization opportunities and in any case open ended waveguides could be found to be too cumbersome for integration in many cases.

Somewhat less work has been reported on dielectric lens designs incorporating amplitude shaping conditions. Here, the motivation is to produce a controlled field distribution over the lens aperture [21]. Routines to calculate two surface lenses for simultaneous phase and amplitude shaping are given in Reference 22. Another motivation is to shape the output beam into an hemispherical or a secant squared (sec^2) pattern for constant flux applications. A renewed interest in these type of patterns comes from emerging millimeter wave mobile broadband cellular systems and wireless local area network applications where non-symmetric lenses may be required [23].

Lens design, based on quasi optics, is addressed either for aperture phase error correction or for output beam shaping. For phase correction the formulation is restricted to circular-symmetric geometries, but methods can be adapted to suit pyramidal dielectric horns as well.

1.3 LUNEBURG AND SPHERICAL LENSES

Spherical lenses will be covered in much more detail in Chapters 6 and 7, along with some accounts of practical developments and applications pursued by the authors in

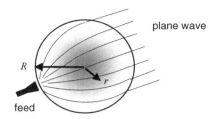

Figure 1.5. Ray paths in a Luneburg lens [24, 25].

the course of several research programs. For the current chapter, a short introduction is offered.

Unlike a paraboloid reflector, or a conventional focusing lens of one of the types introduced in the above paragraphs, a spherical lens does not exhibit a point focus. Rather, its symmetry gives rise to a focal region defined by a spherical surface which is concentric with the lens. Put another way, a feed may be placed at any position around the edge of the lens (or, in special cases, within the interior of the lens). A straightforward corollary of this symmetry is the ability to use several feeds, each giving rise to a separate beam which shares the lens aperture. Thus, a multi-beam antenna is readily offered. Similarly, a feed (or several feeds) mechanically scanned with respect to a spherical lens gives rise to a scanning antenna with very wide scan angle properties, and without scanning loss. A switched beam antenna may similarly be produced.

Where a spherical lens is constructed from a single, homogeneous dielectric material (a polymer e.g., Fig. 1.5) we have a "constant index lens." A disadvantage is that its collimating properties tend to be mediocre, particularly as electrical size increases. In an alternative approach, proposed by R. K. Luneburg in 1943 [24] the sphere is made of materials with non-constant refractive index, that is, where relative dielectric constant ε_r varies with the square of radius:

$$\varepsilon_r = 2 - (r/R)^2$$

where r is radius from the center point and R is outer radius of lens. This formulation gives rise to foci lying on the outer surface at $r = R$. Furthermore, the focus is at a single point in a manner analogous to any properly collimating device (dish, lens etc)— all of the aperture contributes and, given a suitable illumination, the aperture efficiency can be unity at least in theory. This property is irrespective of diameter, quite unlike a constant index lens where the efficiency will be less than unity and also decreases with increasing diameter.

Figure 1.5 illustrates the approximate ray paths for the Luneburg lens case and hence shows curved paths within the dielectric. Luneburg did not have the opportunity to implement such an antenna, as no suitable materials or manufacturing procedures were available at that time. Today, practical "Luneburg" lenses are made from sets of concentric dielectric layers, and as such are really approximations to the ideal case [25].

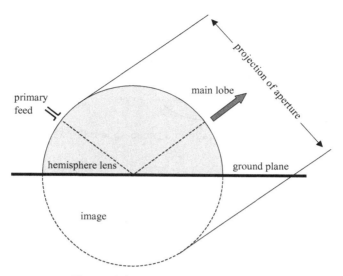

Figure 1.6. Hemisphere lens antenna.

A hemispherical lens antenna is also commonly used in conjunction with a reflective ground plane, the latter gives rise to an image of the hemisphere and so recovers the full aperture that would be presented by a sphere (Fig. 1.6). This arrangement offers mass and space reductions, and can be easier to mechanically stabilize. This might also be called a "lens-reflector" since the flat planar reflector is an integral part of the antenna aperture and contributes to its collimating properties. Of course, the extension of the reflector must be adequate to produce the required image of the hemisphere, and the necessary dimension is also a function of elevation angle (the feed angle with respect to the reflector plane). An inadequate reflector extension will introduce a reduction in effective aperture which amounts to a scanning loss. The topic is covered quite thoroughly in Chapter 7 where recent advances in practical antennas of this type are reported.

This configuration also offers a relatively low-profile solution, and this property makes the hemisphere antenna particularly attractive as a scanning antenna for applications where headroom is limited. The layout is illustrated in Figure 1.7, where it can be seen that the effective aperture height of the hemisphere with reflecting plane can be up to twice that of a conventional reflector antenna [25]. Again, recent practical developments are discussed in Chapter 7.

A variant of the spherical Luneburg lens is cylindrical and where the variation in dielectric constant occurs in just two dimensions rather than three. The "cylindrical Luneburg" lens is therefore somewhat less problematic to manufacture than its spherical parent, but it does offer beam collimation in just one axis, leading to a fan shaped beam rather than a pencil beam. Various approaches to realizing the required variation in dielectric constant, using a juxtaposition of two different materials, are presented in

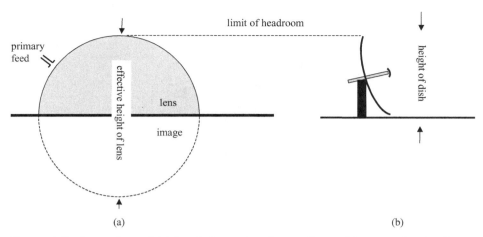

Figure 1.7. Comparison of (a) hemisphere plus plane reflector (b) conventional reflector antenna.

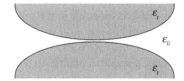

Figure 1.8. Cylindrical Luneburg lens formed from two dielectric discs [26].

Reference 26, along with experimental results. One such geometry is quite elegant because one of the materials is air and the other takes the form of a pair of machined polymer discs as illustrated in cross section in Figure 1.8.

In the 1970s, dielectric lenses were developed in Russia by several organizations [27]; these were mainly used in defense applications. Later, in the 1990s–2000s the Konkur company of Moscow were promoting Luneburg lenses of their own manufacture, although at the time of publication little evidence has been found to indicate that they are still active. Today the Luneburg lens is an attractive candidate antenna for multibeam wideband millimeter wavelength indoor and outdoor communication systems and for airborne surveillance radar applications. Manufacturers that appear to be still active are "Luntec" of France, and Rozendal Associates of Santee, California. Also, during the 2000s, Sumitomo Electric Industries of Japan were manufacturing lenses for multi-beam satellite TV, primarily in the receive only domestic market in Japan. (At the time of writing it is not known to the authors whether this continues although some anecdotal evidence suggests not. The brand "LuneQ" is still encountered in this context.)

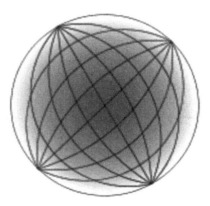

Figure 1.9. Maxwell's fish-eye lens, with dark color representing increasing refractive index.

1.4 QUASI OPTICS AND LENS ANTENNAS

Originally, quasi-optics (QO) applications of classical homogeneous lens designs remained restricted to aperture phase correction in millimeter wave horn feeds. At the same time, inhomogeneous lenses that were only theoretical curiosities at traditional optical wavelengths were realized in the microwave range and still attract much attention. These include the Luneburg lens, the Maxwell "fish-eye" lens (Fig. 1.9), and others. The spherical or cylindrical Luneburg lens (as we have seen) has a dielectric constant varying smoothly from 2 at its center to 1 at its outer boundary, which is the focal surface in the geometric optics (GO) approximation. While lens dimensions are larger than microwave wavelengths, spheres with a dielectric constant varying smoothly on the scale of the wavelength are almost impossible to fabricate. Therefore various sorts of Luneburg lenses which employ discrete dielectric layers have been devised. The oldest types consisted of a finite number of spherical or cylindrical layers each with constant permittivity.

Quasi optics can be considered to be a specific branch of microwave science and engineering [28–30]. The term "quasi optics" is used to characterize methods and tools devised for handling, both in theory and in practice, electromagnetic waves propagating in the form of directive beams, width w is greater than the wavelength λ, but which is smaller than the cross-section size, D, of the limiting apertures and guiding structures: $\lambda < w < D$.

Normally we have $D < 100\lambda$, and devices as small as $D = 3\lambda$ can be analyzed with some success using QO. Therefore QO phenomena and devices cannot be characterized with geometrical optics (GO) that requires $D > 1000\lambda$, and both diffraction and ray-like optical phenomena must be taken into account. It is also clear that, as Maxwell's equations (although not material equations) are scalable in terms of the ratio D/λ, the range of parameters satisfying the above definition sweeps across all the ranges of the electromagnetic spectrum, from radio waves to visible light (Fig. 1.10) and beyond.

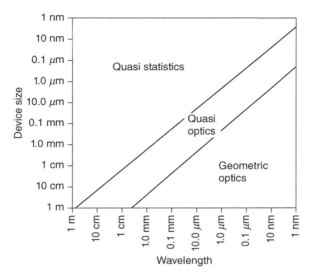

Figure 1.10. A diagram showing the place of quasioptical (QO) techniques with respect to geometrical optic (GO) and quasistatic techniques, in the plane of the two parameters-device size and wavelength.

Therefore QO effects, principles and devices can be encountered in any of these ranges, from skyscraper-high deep-space communication reflectors to micron-size lasers with oxide windows.

As a universal QO device, the dielectric lens was first borrowed from optics by O. Lodge for his experiments at a wavelength of 1 meter in 1889 [31], then used in microwave and millimeter wave systems in the 1950–1980s, and is today experiencing a third generation in terahertz receivers. Moreover, as the above relation among the device size, beam size, and wavelength is common in today's optoelectronics, it is clear that QO principles potentially may have a great impact on this field of science as well. Nevertheless, in the narrow sense, the term QO still relates well to the devices and systems working with millimeter and sub-millimeter waves. E. Karplus apparently coined the term quasi optics in 1931 [32], and then it was forgotten for exactly 30 years before being used again [33]. A parallel term *microwave optics* was used in several remarkable books and review articles of the 1950–1960s [34].

If compared with the classical optics of light, millimeter wave and sub- millimeter wave QO have several features. First, electromagnetic waves display their coherence and definite polarization state. Also they display much greater divergence and diffraction, while direct measurements of their amplitude and phase are relatively easy.

We then give a QO example to shape the horn aperture into a lens. Consider the geometry of Figure 1.11, which represents an axis-symmetric homogeneous solid dielectric horn with shaped aperture. Starting from the spherical wave front inside the dielectric horn, we define associated rays originating at point Q and refracting at surface $r(\theta)$ according to Snell's laws.

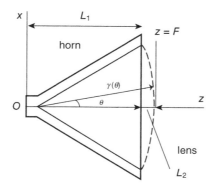

Figure 1.11. Geometry of axis-symmetric solid dielectric horn with shaped aperture for phase correction.

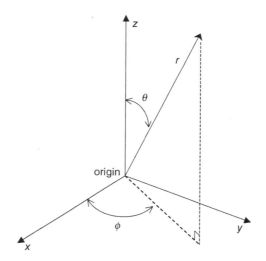

Figure 1.12. Spherical coordinate system.

The optical path along the rays from the origin O to plane $z = F = L_1 + L_2$ must be constant for all elevation angles θ and azimuth angles φ (Figs. 1.11 and 1.12)

$$r(\theta)\beta + F - r(\theta)\cos\theta = F\beta \tag{1.1}$$

Where r is the radius of the horn which is subject to the value of θ, and β is the normalized longitudinal propagation constant.

Equation (1.1) is written as if β were constant with z. Actually β changes with z but only from lens shape, in a region where the horn cross section dimensions are expected to be much larger than λ where β is almost constant and approaches $\varepsilon^{0.5}$. Under this assumption, rearranging (Eq. 1.1) we obtain

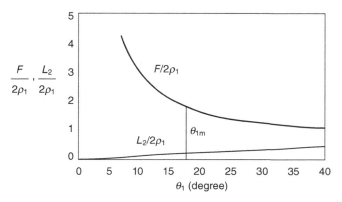

Figure 1.13. Lens depth and lens-horn depth to aperture ratio versus horn semi-fare angle θ for polystyrene material with dielectric constant ε = 2.5.

$$r(\theta) = \frac{F(\sqrt{\varepsilon} - 1)}{\sqrt{\varepsilon} - \cos\theta} \tag{1.2}$$

It's helpful to then express F in terms of the horn semi-flare angle θ_1 and the aperture radius ρ:

$$F = \frac{\rho}{\sin\theta_1}\left[\frac{\sqrt{\varepsilon} - \cos\theta_1}{\sqrt{\varepsilon} - 1}\right] \tag{1.3}$$

Figure 1.13 shows the lens depth and lens-horn depth to aperture diameter ratio $L_2/2\rho_l$ and $F/2\rho_l$ versus horn semiflare angle θ_1. For the same flare angle $L_2/2\rho_l$ and $F/2\rho_l$ are larger for smaller values of ε. The ratio $L_2/2\rho_l$ grows faster with θ. The ratio $F/2\rho_l$ becomes more favorable as θ_1 increases, but there is an upper limit for this angle for two reasons:

Firstly, θ_1 must not exceed the value that allows the propagation of higher-order modes in the dielectric horn.

Secondly, the lens depth should not exceed the ellipse major semiaxis, that is

$$\theta_{1m} \leq \mathbf{arc}\cos\left[\frac{1}{\sqrt{\varepsilon}}\right] \tag{1.4}$$

For polystyrene material (ε = 2.5) we obtain θ = 51 degrees, far beyond the single mode operation condition.

The analysis and design of QO components could either lose accuracy in the range of the characteristic QO relationship between the wavelength and the size of the scatterer, or lead to high numerical complexity of the algorithms and hence prohibitively large computation times. Therefore, one would look in vain for the frequency dependence of the gain of a reflector or lens antenna of realistic size computed with the method of moments or Finite-difference time-domain method (FDTD).

Exciting opportunities for the design of revolutionary new QO components and instruments are offered by emergent technological innovations, such as electromagnetic bandgap materials and meta-materials (also known as twice-negative and left-handed materials). For example, a QO prism made of bandgap material may display frequency and angular dependence of the incident beam deflection one hundred times stronger than that of a similar homogeneous prism [35]. Exotic designs of new QO lens antennas and beam waveguides made of meta-materials can be based on interesting effects including negative refraction [36]. Therefore the future of quasi optics is perfectly secure as long as electromagnetic waves are still used by the information society.

1.5 LENS ANTENNA DESIGN

The usual starting point for lens antenna design is to apply geometric optics. Here, a ray tracing approach is used, where the radiation is modeled as rays which radiate from an common origin, or source. Incidentally, this origin would be quite analogous to the antenna phase center. Of course, the validity of this simple approach is questionable because the lens typically lies within the near field region of the source or primary feed. GO is nevertheless though usually thought a valid starting point, particularly for predicting the properties of the main lobe wherein most of the radiation is contained. A quite surprisingly good agreement will be observed with measurement results in this main lobe regime. The sidelobes region, contrastingly, is not likely to be described with much accuracy.

Next, it is also simplest to design for lens cross sections which are circular (though they need not be). It then follows that the primary feed radiation pattern is circularly symmetric, though, again, it need not be. These starting assumptions removed dependence upon angle ϕ, the angle of rotation about the antenna axis of symmetry or boresight.

In principle though, the lens surface may be designed to accommodate asymmetries in the primary feed pattern, or ϕ angle dependency. These might be corrections for amplitude and also phase. In a more extreme case, a highly elliptical primary pattern (very different beamwidths in orthogonal planes, such as E-plane and H-plane) would require a lens cross section with similar ellipticity, although this seems a rare design objective and we present no examples. On the other hand, an antenna with a highly symmetric feed pattern but asymmetric secondary pattern is described in Chapter 5. This has a circular cross section, but still a ϕ dependency of the lens profile.

Returning to GO, its first principle is that the rays trace the flow of power between points. It follows from this that the aggregate flow of power through a closed surface comprises a bundle of such rays. Through such a bundle, or tube, power flow is constant across any cross section. The second pillar of GO is Fermat's principle which states that a ray's path is that which is of shortest length (or time) between any two given points. This underpins the optics of mirrors, reflection and refraction, and from which Snell's law may be derived.

A generalized design procedure for lenses can be listed as the following steps.

(i)   Clarify the design objective for beamwidth and gain—this sets the required electrical dimensions.
(ii)   Choose the operation frequency and bandwidth—the former sets physical dimensions.
(iii) Choose a suitable material (see Section 1.1.3)
(iv)   Estimate the loss in the material. From this, dimensions may need to adjusted to realize the required gain.
(v)   Determine the primary feed type, or other method of illumination.
(vi)   Iterations as often as needed for fine-tuning the antenna performance.

For a lens which is properly designed, and acts as a collimating aperture, the directivity is determined by the aperture area. The gain will be lower than directivity by the sum of the various loss terms, including conductor loss, dielectric loss, and feed spillover loss (that proportion of the feed's radiated power which is not incident upon the lens surface). On aperture area, a simple rule of thumb is that doubling the aperture diameter will increase the gain by 6 dB, since the area of the aperture quadruples. For instance, an 12 cm lens fed by a 6 cm horn would add 6 dB to the gain of the horn, and a 24 cm lens would add 12 dB. Modest gain improvements take modest sizes, but beyond this big gain increases lead to physically large antennas and so for lenses working at the lower microwave frequencies the mass could become prohibitive. Of course, a 6 dB increase in gain will double the range of a communications system over a line-of-sight path.

Feed horn dimensions may be designed for the intended application, but in some cases it may be more practicable to use a commercial item.

If we start with a given primary feed, the beamwidth of the horn (usually smaller than the physical flare angle of the horn) will strongly influence the focal length of the lens. Looked at from the other point of view, if the focal length of the lens is the starting point, this determines the necessary feed properties. Clearly, longer focal lengths place the feed at a greater distance and so need a narrower beamwidth horn.

Approximations for the 3 dB beamwidths (respectively $\theta_{3dB}E$ and $\theta_{3dB}H$ for the E-plane and H-plane) often encountered in antenna theory are:

$$\theta_{3dB}E \approx \frac{57}{L_{E\lambda}} \text{ degrees}$$

$$\theta_{3dB}H \approx \frac{68}{L_{H\lambda}} \text{ degrees}$$

where $L_{E\lambda}$ is the aperture dimension in wavelengths along the E-plane direction, and $L_{H\lambda}$ is the aperture dimension in wavelengths along the H-plane direction. These relations reflect the well known inverse law between beamwidth dimension. Often, different numbers are used in the numerator on the right hand side, depending on assumptions about the illumination edge taper. A stronger taper, or roll-off at the aperture limit, produces a wider beam. The above figures are typical for a pyramid type horn, and the wider H-plane beamwidth reflects the tendency for field distribution of the waveguide TE_{10} mode to decay along this direction.

Considering the shape of the lens, this depends on the refractive index n (the ratio of the phase velocity of propagation of a radio wave in a vacuum to that in the lens). A slow-wave lens antenna, as in optics, is one for which $n > 1$. A fast-wave lens antenna is one for which $n < 1$ but this does not have an optical analogy, at least for classical materials. Considering the Luneburg lens once again as an example, each point on its surface is the focal point for parallel radiation incident on the opposite side. Ideally, the dielectric constant ε_r of the material composing the lens falls from 2 at its center to 1 at its surface (or equivalently, the refractive index n falls from $\sqrt{2}$ to 1, according to

$$n = \sqrt{\varepsilon_r} = \sqrt{2 - \left(\frac{r}{R}\right)^2}$$

where R is the radius of the lens. It should be added here that because the refractive index at the surface is (in theory at least) the same as that of the surrounding medium, no reflection occurs at the surface. In practice of course there must be a transition at a boundary between a real material with $n > 1$ and air with $n = 1$, but the closer to unity that the outer index can be made, the lower the reflection loss. In any case, the effect is sometimes over-estimated in spherical microwave lenses and the topic is taken up again in Chapter 6.

Returning to lens design in general we next calculate the lens curvature. In one approach the equation for lens curvature can be derived where there is just one refracting surface. The other surface is planar, and since rays enter it at normal incidence is does not produce refraction (Fig. 1.14). As noted above, we are now assuming rotational symmetry: there is no ϕ dependence and the problem reduces to two dimensions. The surface is most easily understood by equating all possible path lengths, including that along the central axis. In Figure 1.14, an arbitrary path has length r in air and l in the dielectric, while the axial path's components are the focal distance F in air and the maximum lens thickness T in the dielectric. Equating the electrical lengths yields:

$$nl + r = nT + F$$

This is also expressed in polar co-ordinates (r, θ) as

$$r = \frac{(n-1)F}{n\cos\theta - 1} \tag{1.5}$$

This equation defines a hyperbolic curve with eccentricity of n and where the origin coincides with the focus. Later, the profiles of other types of lenses will be described, for example, with two refractive surfaces. The design procedures follow a similar approach as for this simple case.

For collimation the optimum curvature is elliptical, but because we also know that a spherical curvature is commonly encountered in optics, we can say that a circle should

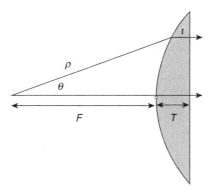

Figure 1.14. One-surface-refracting lens.

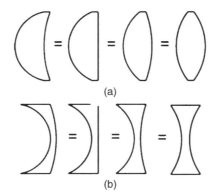

Figure 1.15. (a) equivalent optical lenses (b) equivalent metal lens antennas. All have the same focal length.

be a usable approximation to the planar-concave geometry of Figure 1.14. We can show that the circle is indeed a good fit if the focal length is more than twice the lens diameter, that is, it is an $f/2$ lens. From this geometry it can be shown that the beamwidth of the feed shouldn't exceed 28 degrees, requiring its aperture dimension to be 2 wavelengths or greater.

The radius of curvature R of the two lens surfaces is calculated from an optical formula:

$$\frac{1}{f} = (n-1)\left(\frac{1}{R_1} - \frac{1}{R_2}\right)$$

where a negative radius denotes a concave surface. All combinations of R_1 and R_2 which satisfy the formula are equivalent, as shown in Figure 1.15.

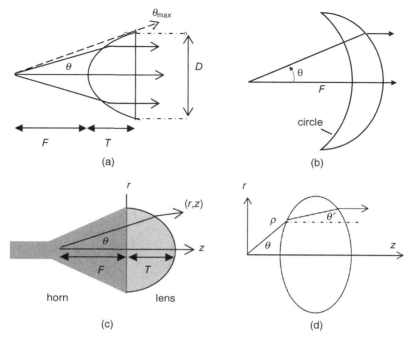

Figure 1.16. Types of lens: (a) type *a*; (b) type *b*; (c) type *c*; (d) type *d*.

The over-arching objective of the collimating lens, like that of the parabolic reflector, is an aperture exhibiting a constant phase front. To this end, the four most common lens types will be reviewed, as shown in Figure 1.16. Types *a* and *b* have one refracting surface, and one where rays pass at normal incidence. We have already encountered type *a* above and in Figure 1.14. In contrast, in type *b*, it is the surface on the opposite side from the feed at which refraction takes place, and the facing side is spherical. Types *c* and *d* lenses are often called "dual-surface lenses" because both are used to produce a required amount of refraction which together produce collimation and the required constant phase front. Type *d* has two similar refracting surfaces while Type *c* has two dissimilar surfaces (one is plane), but can be identified as a sub-class of type *d* and as one which is favorable for placing directly in the aperture of a horn, as illustrated.

Two formulations are now useful for describing each type of lens. These are a formula for the lens thickness, and another for the surface, although they are really equivalents. The profile for the type *a* lens has already been covered above. Its thickness *T*, relative to its diameter *D*, is given by:

$$\frac{T}{D} = \sqrt{\left\{ \frac{1}{n^2}\left(\frac{F}{D}\right)^2 + \frac{1}{4n(n+1)} \right\}} - \frac{F}{nD}$$

Moving on to the type *b* lens, the surface facing the feed is spherical, while the opposite face has a profile described by the polar equation:

$$r = \frac{(n-1)}{\cos\theta + n}F$$

and thickness:

$$\frac{T}{D} = \frac{1}{2\sqrt{(n^2 - 1)}}$$

If the type a lens is intended to be placed in the aperture of a horn, and in contact with it, the angle θ_{max}—Fig. 1.16a—must exceed the horn's interior angle of taper, where

$$\theta_{max} = \cos^{-1}\left(\frac{1}{n}\right)$$

If this condition isn't met, a physical contact can't be made across the full aperture of the horn. This need not necessarily be a problem, but would require an additional support structure between the horn and lens, a longer structure, and could give rise to excessive spill-over radiation unless the horn's (primary feed) pattern is carefully chosen.

For the type *c* lens, as a matter of convenience, we use a radius *r* along the transverse axis which is orthogonal to the usual lens axis *z*. The shaped lens profile is then given by *r* and *z*, both expressed as functions of angle θ:

$$r = \frac{(n-1)T + (n^2 - 1)F\sec\theta - FS\tan\theta}{\left(\dfrac{n^2}{\sin\theta} - S\right)} \tag{1.6}$$

and

$$z = (r - F\tan\theta)S \tag{1.7}$$

where

$$S = \sqrt{\left\{\left(\frac{n}{\sin\theta}\right)^2 - 1\right\}}$$

The axial thickness of the type *c* lens, again relative to diameter *D*, is

$$\frac{T}{D} = \left[\sqrt{\left\{1 + \frac{1}{\left(\dfrac{2F}{D}\right)^2}\right\}} - 1\right]\frac{F}{\dfrac{D}{n-1}} \tag{1.8}$$

The equations defining the surfaces of type d lenses cannot be expressed as directly as with the preceding 3 cases. Instead, they are related through simultaneous differential equations. These have been expressed by Olver et al. [18] as:

$$\frac{d\rho}{d\theta} = \frac{n\rho \sin(\theta - \theta')}{n\cos(\theta - \theta') - 1}$$

$$\frac{dr}{d\theta'} = \frac{P(\theta')\sin\theta' r_{max}^2}{2r \int\limits_{0}^{\theta'_{max}} \rho(\theta')\sin\theta' d\theta'}$$

$$\frac{dz}{dr} = -\tan\left\{\theta' + \tan^{-1}\left(\frac{\sin\theta'}{n\cos\theta'}\right)\right\}$$

where $P(\theta')$ is the radiated power per unit solid angle, and in the direction:

$$\theta' = \theta - \frac{\pi}{2} + \tan^{-1}\left(\rho\frac{d\theta}{d\rho}\right) + \sin^{-1}\frac{1}{n\sqrt{\left\{1 + \left(\rho\frac{d\theta}{d\rho}\right)^2\right\}}}$$

The axial thickness T of a type d lens is given [18] by:

$$\frac{T}{D} = \sqrt{\left\{\frac{1}{n^2}\left(\frac{F}{D}\right)^2 + \frac{1}{4n(n-1)}\right\}} - \frac{F}{nD}$$

Unsurprisingly, the type d lens, as well as being of more complex design, entails more fabrication stages compared to its type c counterpart, and possibly offers little, if any, performance advantage.

The relative permittivity of the lens material shouldn't be less than a certain minimum value if the primary feed is intended to be a horn type in contact with the lens. In this context, a long horn with a narrow flare angle would be needed with very low constant lenses which of necessity are physically thick (these fall into the low density foam category with $\varepsilon_r < 1.5$). Such a horn-lens combination may offer minimal advantage compared to the horn on its own, and a better combination would use a higher relative permittivity, for example, >1.7. More details can be found in Reference 37.

One of the main disadvantages of dielectric lenses is their bulk at lower microwave frequencies. To overcome this, the thickness of a solid-dielectric lens can be reduced by removing slabs of integral-wavelength thickness at periodic intervals called *zones*. Applying this to a type c lens gives rise to the profile shown in Figure 1.17. Here, a point at transverse radius r and axial distance z is shown.

When the zoning principle is applied across the lens surface the optical-path length differs by one wavelength between adjacent zones. Starting from the center and increasing radial distance r, the next zone interval is the point at which the lens thickness has reduced by one wavelength. Hence the next zone commences with an increase in

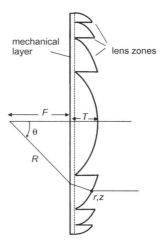

Figure 1.17. Zoned lens.

thickness of the same amount, which is $\lambda_0/(n-1)$. The minimum physical lens thickness needs to be at least a little larger than this value to provide a mechanical support layer, as indicated on the left side in Figure 1.17. For N zones, the path through the outer zone is $\lambda_0(N-1)$. The zoned lens is hence a limited bandwidth device, and becomes more so as its diameter increases and more zones are used. Hence this approach detracts from what is otherwise an inherently wideband device.

From Figure 1.17, the equations for the profile in terms of both angle θ and R are just modified versions of those encountered for the continuous surface type c lens, and are:

$$r = R\sin\theta + \frac{z}{S}$$

and

$$z = \frac{(N-1)\lambda + (n-1)T + F - R}{\dfrac{1}{S}\left(\dfrac{n}{\sin\theta}\right)^2 - 1}$$

where $N = 1,2,3. \ldots$ etc counting from the lens center, and S is the same as that in (Eq. 1.8).

Zoned lenses are best used at lower microwave frequencies rather than at millimeter wave frequencies. In the former case the mass saving can be significant and it is worth the extra fabrication complexity, so long as bandwidth requirements are not great. A subtle disadvantage of zoning is that aperture blockage is introduced by the discontinuities between zones, an effect which is a strong function of f/D value [38].

Another development of the zoned lens is the phase correcting Fresnel zone plate lens (Fig. 1.18). Here the manufacturing process is simplified because the curvature of

s

Figure 1.18. Phase Correcting Fresnel Zone Plate Lens.

a conventional lens (zoned or otherwise) is dispensed with and a set of radial grooves at discrete depths used instead. The radii of the grooves are given by:

$$r_i = \sqrt{2Fi\frac{\lambda_0}{p} + i\left(\frac{\lambda_0}{p}\right)^2} \quad i = 1, 2 \dots$$

and the thickness of the step s by:

$$s = \frac{\lambda_0}{p\left(\sqrt{\varepsilon_r} - 1\right)}$$

where F is focal length, λ_0 the free space wavelength and p is an integer which defines the number of steps and hence the granularity of the surface. For example, for $p = 2$ there are 2 corrections per wavelength or a $180°$ phase correction, for $p = 4$ there are 4 corrections per wavelength or a $90°$ phase correction and so on. (For $p = $ infinity the steps are blended out into a continuous curved surface.)

Metal plate lenses, encountered above in Figure 1.2, have been used as a potentially less massive alternative to either smooth or zoned dielectric lenses, with a different set of strengths and weaknesses in degrees of freedom for the design—among these being bandwidth and polarization.

1.6 METAMATERIAL LENS

Metamaterials are artificial composite structures with artificial elements (much smaller than the wavelength of electromagnetic propagation) situated within a carrier medium. The subject attracted significant interest since practical implementation solutions have emerged. Also, research work shows that antenna gain can be enhanced by using metamaterials as antenna substrates [39].

These materials can be designed with arbitrary permeability and permittivity [40]. Left-handed materials are characterized by a negative permittivity and a negative permeability- at least across a portion of the electromagnetic frequency spectrum. As a consequence, the refractive index of a metamaterial can also be negative across that portion of the spectrum. In practical terms, materials with a negative index of refraction are capable of refracting propagating electromagnetic waves incident upon the metamaterial in a direction opposite to that of the case where the wave was incident upon a material having a positive index of refraction (the inverse of Snell's law of refraction in optics). If the wavelength of the electromagnetic energy is relatively large compared to the individual structural elements of the metamaterial, then the electromagnetic energy will respond as if the metamaterial is actually a homogeneous material.

As these materials can exhibit phase and group velocities of opposite signs and a negative refractive index in certain frequency ranges, both of these characteristics offer a new design concept for lens antenna feeding.

One of the approaches starts from the equivalent transmission line model and artificially loads a host line with a dual periodic structure consisting of series capacitors and shunt inductors [41]. The length of the period and the values of the capacitors and inductors determine the frequency band in which the material has this double negative behavior.

An example of such structures is in arrays of wires and split-ring resonators [42, 43]. These three-dimensional structures are complicated and are difficult to apply to RF and microwave circuits. A more practical implementation uses transmission lines periodically loaded with lumped element networks [44, 45].

The starting point is the transmission line model presented in Figure 1.19a. The equivalence between the distributed L and C for the transmission line and the permittivity and permeability of the medium is expressed as $\varepsilon = C$, $\mu = L$. By periodically loading this transmission line with its dual in Figure 1.19b, the values of ε and μ change as follows [44]:

$$\varepsilon_{eff} = \varepsilon - \frac{1}{\omega^2 Ld} \quad \mu_{eff} = \mu - \frac{1}{\omega^2 Cd} \tag{1.9}$$

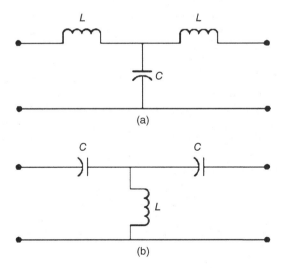

Figure 1.19. (a) *L-C-L* and (b) *C-L-C* transmission line models.

Interdigital capacitors

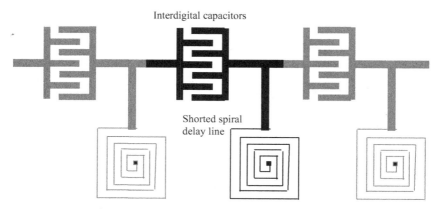

Shorted spiral
delay line

Figure 1.20. Periodic structure of three unit cells.

where ε and μ are the distributed inductance and capacitance of the host transmission line respectively. It is clear from Equation (1.9) that for certain values of L, C and d, the effective permittivity and permeability of the medium becomes negative for some frequency ranges. In these ranges, the refractive index is negative and the phase and group velocities have opposite signs.

In Figure 1.20, three unit-cell circuit structures are repeated periodically along the microstrip line. A unit-cell circuit, in the structure, consists of one or more electrical components that are repeated—in this case disposed along the microstrip transmission line. In the structure in Figure 1.20 above, series interdigital capacitors are placed periodically along the line and T-junctions between each of the capacitors connect the microstrip line to shorted spiral stub delay lines that are, in turn, connected to ground by vias. The microstrip structure of one capacitor, one spiral inductor and the associated ground via, form the unit-cell circuit structure of Figure 1.20.

Structures such as Figure 1.20 can be used in leaky-wave antennas (as opposed to phased array antennas), which have been designed to operate at frequencies of up to approximately 6.0 GHz [46]. With certain modifications, these metamaterials can be used at relatively high frequencies, such as those frequencies useful in millimeter wave communications applications [47]. For instance, the unit-cell circuit structure of Figure 1.20 can be reduced to a size much smaller than the effective wavelength of the signal. To achieve a high-performance transmission line impedance at a particular frequency, the physical size and positioning of unit cells in the metamaterial microstrip line needs to be carefully considered.

High-gain printed arrays have previously relied on a signal-feed/delay line architecture that resulted in a biconvex, or *Fresnel lens* for focusing the microwaves [47]. The use of such lens architectures has resulted in microwave radiation patterns having relatively poor sidelobe performance due to attenuation as the wave passed through the lens. Specifically, the signal passing through the central portion of the lens tended to be attenuated to a greater degree than the signal passing through the edges of the lens.

This resulted in an aperture distribution function that was "darker" in the center of the aperture and "brighter" near the edges. The diffraction pattern of this function results in significant sidelobes (the diffraction or far-field radiation pattern is the two-dimensional Fourier transform of the aperture distribution function). While placing signal delay lines in the lens portion of the system could reduce the sidelobes and, as a result, increase the performance of a phased-array system, this was deemed to be limited in its usefulness because, by including such delay lines, the operating bandwidth of the phased-array system was reduced.

However, instead of a biconvex lens, a metamaterial can be used to create a biconcave lens (by means of controlling the effective refractive index of the material) for focusing the wave transmitted by the antenna. As a result, a wave passing through the center of the lens is attenuated to a lesser degree relative to the edges of the lens (the aperture is now brighter at the center and darker near the edges), thus significantly reducing the amplitude of the sidelobes of the antenna while, at the same time, retaining a relatively wide useful bandwidth.

The metamaterial lens is used as an efficient coupler to the external radiation, focusing radiation along or from a microstrip transmission line into transmitting and receiving components. Hence, it can be designed as an input device. In addition, it can enhance the amplitude of evanescent waves, as well as correct the phase of propagating waves.

A meta-material lens is made using composite right/left handed transmission line (CRLH-TL) [48]. The potential of the shaped metamaterial lens has been investigated for wide angle beam scanning [49]. One example is a meta-material lens antenna using dielectric resonators for wide angle beam scanning. The lens antenna is composed of the radiator, the parallel plate waveguide and the meta-material lens.

Figure 1.21 shows an unit cell of the meta-material lens. The unit cell is composed of the parallel plate waveguide with $\varepsilon_r = 2.2$ and the dielectric resonator with $\varepsilon_r = 38$, thickness is $h = 2.03$ mm and the diameter is $a = 5.1$ mm. The distance of the parallel

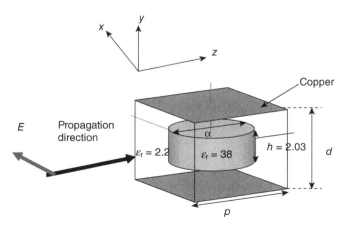

Figure 1.21. Configuration of unit cell of the metamaterial lens (©2010 IEEE [49]).

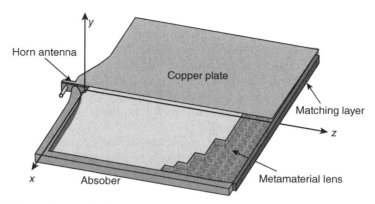

Figure 1.22. Configuration of the metamaterial lens antenna for wide angle beam scanning (©2010 IEEE [49]).

plate is d. The excitation mode of the parallel plate waveguide is TE_1 mode in order to excite $TE_{01\delta}$ mode of the disk type dielectric resonator. The effective permittivity ε_{eff} and permeability μ_{eff} are shown in Figure 1.22.

The frequency at $\varepsilon_{eff} = 0$ increases as d decreases because the cut-off frequency shifts to a higher frequency. When $d = 5.1$ mm, negative values of ε_{eff} and μ_{eff} are obtained below 11.5 GHz and the refractive index n becomes negative value. Here, $n = -1$ is obtained around at 11 GHz. Figure 1.22 shows the configuration of the meta-material lens antenna composed of the radiator, the parallel plate waveguide and the metamaterial lens. The number of unit cells is 100, $f/D = 0.5$ and the lens aperture is 156 mm (about 5.7 λ). The minimum thickness of the metamaterial lens along z axis is three cells. The matching layer is set on the aperture of the lens for the impedance matching between the metamaterial lens and the air. The absorber materials are set around the lens antenna for reduction of the multiple reflections in the parallel plate waveguide. The parallel plate waveguide between the radiator and the metamaterial lens is filled by dielectric material with $\varepsilon_r = 10.2$.

1.7 PLANAR LENS OR PHASE-SHIFTING SURFACE

A concept gaining ground since the mid 1990s has been the planar lens [50] or phase-shifting collimating surface. Like the reflect-array, the planar lens comprises a surface of discrete radiating elements configured to yield a phase distribution which exhibits a collimating effect. While the origin of the "reflect-array" moniker would be self-evident, the equivalent "lens-array" appears not to have come widely into use. Notwithstanding this nomenclature, the two techniques are closely related, just as the reflector and lens antennas are related: each exhibits a first focus at which a primary feed is placed, and a second focus at an infinite distance.

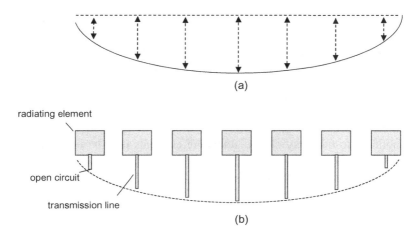

Figure 1.23. Reflect-array concept: (a) path lengths across aperture of parabolic reflector, and (b) discretized equivalent using radiating elements and transmission lines.

1.7.1 Reflect Array

The reflect-array has received rather more attention than its lens-like equivalent and so it's worth proceeding with a very short review of this technology. The principle of operation of the planar lens should then be more easily grasped.

The essential property of the reflect array is to re-radiate the incident field of a source, or primary feed, in a manner which yields a constant-phase aperture and thus a collimated beam. While collimation is achieved by a parabolic reflector owing to its curvature, the reflect array utilizes a discrete set of array elements, distributed across a planar aperture, whose phase responses are suitably tailored. The two cases are compared in Figure 1.23a,b. While a linear array is shown the same phase shifting principle could apply also across two dimensions. The radiating elements could be any convenient antenna (in some early reflect arrays waveguide horns were used) although a popular and practical type would be a printed dipole, patch or similar. A means of varying the phase response of each radiator (i.e., the phase of the reflected field compared to the incident field), and one which is perhaps easiest to visualize, is the series addition of a short or open-circuited transmission line. Then, varying the length of this line varies the phase of the element's re-radiated field.

Considering printed radiating elements it is easy to see how these can conveniently be integrated with printed microstrip transmission lines whose lengths are chosen to produce the required aperture phase distribution, albeit in a discretized manner. Now, open circuited microstrip lines exhibit some disadvantages: consumption of circuit real estate, a tendency to produce unhelpful radiation at the open circuit stub, and circuit loss. These drawbacks can to some extent be ameliorated by use of aperture coupled patches so that parasitic radiation can be suppressed on the opposite surface to the radiating aperture. Alternatively, the microstrip lines can be dispensed with altogether if alteration of some other dimension (e.g., patch length and width) can be contrived to generate an adequate variation in phase.

A cited advantage of reflect arrays includes that of fabricating a reflecting, collimating surface without recourse to the perceived manufacturing cost of fabricating a parabolic dish. Now, while dishes up to a certain size (say 0.5 to 1.0 m, as used for satellite TV) are nowadays remarkably low in cost, other properties of reflect arrays compared to parabolic reflectors should also be set out. The reflect array allows for further manipulation of the aperture distribution, for example by integrating active phase shifters or amplifiers to the array circuit. Radiating elements may be tuned to particular frequencies or to introduce polarizing effects (e.g., yield circular polarization from a linear polarized feed). In Reference 51 a switched beam variant was reported, whereby the phase distribution across the aperture was altered by mechanical adjustment of one of the laminates.

Furthermore, beam shaping can be achieved by imposition of an arbitrary phase and amplitude distribution across the array, and this can be achieved at low cost and for low production volumes should the array be fabricated using printed circuit technology. The reflect array has also established interest as a stowable or foldable antenna; this has been of particular interest in space communications where large apertures might be stowed in relatively small volumes for launch, then unfurled for deployment in orbit. Some remarkable examples of experimental antennas for space have been inflatable reflect arrays developed at Jet Propulsion Laboratory [52, 53]. Reflect arrays of course are not without disadvantages. Perhaps the most prominent among these are the inherent low bandwidth which is caused by the frequency dependence of delay-lines (if used) or other phase altering property of the elemental radiators.

Other types of reflect array may be illuminated from a distance and so both foci lie at infinity. These may also phase modulated so as to impart a unique code on the reflected signal [54] although these reflectors act as transponders rather than antennas in the present context. One such device is shown in Figure 1.24 where the radiating

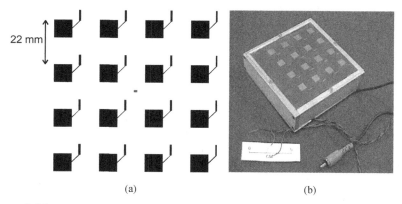

(a)   (b)

Figure 1.24. An active reflect-array (a) circuit for x-band, (b) assembled transponder.

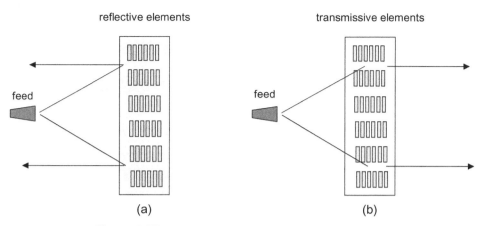

Figure 1.25. Reflect array (a) compared to lens array (b).

elements are visible and the active elements are housed on a second laminate which is aperture-coupled to the radiating layer.

1.7.2   Planar Lens or Lens Array

Having introduced the basic properties of the quite well established and studied reflect array, the principle of the planar lens, or as has been coined here "lens array," should be apparent. The terms "discrete lens" and "transmit array" have also been used. Where the radiating elements of the reflect array are replaced with transmissive elements (Fig. 1.25), and so long as elemental phase distribution is retained, the surface resembles a lens rather than reflector. Put another way, the radiating elements are now required to forward scatter rather than back scatter. In practice, this would likely be achieved by dispensing with the ground plane associated with the reflect-array's patch antennas, and optimizing the forward scattering efficiency of the elements. This would lead to a single layer lens array. Suitable radiating elements could then closely resemble those commonly encountered in frequency selective surfaces, for example, loop, star, slot, ring, or dipole types [55]. A difference however is that the elements of the single layer lens array will not be of uniform dimension, since the phase of the forward scattered component must be a function of the path distance to the primary feed. Ideally, it should be possible to set the phase response of each element within the range 0° to 360° which is not always straightforward depending on the type of radiating element chosen. Another type of radiator reported is the stacked (multi-layer) patch, where several degrees of freedom are available to help tune to the desired phase response. In Reference 56 three metallic and two dielectric layers were used to realize phase shifting surfaces at 30 GHz.

Alternatively, a dual-layer lens array can be used where one array acts as a receiving layer and each element is connected to a counterpart in a second, transmitting array.

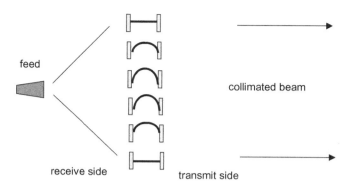

Figure 1.26. Two layer lens array with interconnecting lines which determine phase.

The lengths of the interconnecting lines determine the required phase relationship between transmitting and receiving elements according to their location in the array planes (Fig. 1.26). Here it is more straightforward to set the phase relationship between the two interconnected elements: the length of the transmission line is increased or decreased as necessary, although this does lead to the disadvantages of ohmic loss and the consumption of circuit real estate.

A subtle effect that needs to be considered in lens arrays is that both the element frequency response and the phase of its transfer function tend to vary with the incident angle of the incident wave (the incidence angle with respect to the feed), particularly for short f/D ratios. For f/D much greater than 1 the variation of incidence isn't very great and its effect might justifiably be ignored.

In Reference 57 a variant was reported whereby the intermediate layer between receive and transmit printed antennas comprised a co-planar-waveguide resonator. The properties of this resonator determined the phase relationship between the two antennas and so acted in lieu of the more simplistic transmission line interconnection illustrated above. Furthermore, the resonator exhibited filter properties and so the term "antenna-filter-antenna" was coined for this frequency selective radiating element. Phase ranges of only 0° to 180° were achieved for a given antenna element type, but the full 0° to 360° range was achieved by combining two different types across the planar lens aperture. Abbaspour-Tamijani et al. [57] reported an experimental 3-in.- diameter lens array with measured gain 25.6 dBi at 35.3 GHz and so with an effective insertion loss of 3.5 dB.

Another type of planar lens antenna is the *Fresnel zone-plane lens* (Fig. 1.27). The origin of this lens is from its optical equivalent where a flat lens was sought. Its principle of operation is that adjacent rings have a mean phase difference of $\pi/2$- these are Fresnel zones. By blocking out alternate zones, those remaining zones have in-phase average path lengths. Like the zoned dielectric lens, the zone-plate is bandwidth limited because its design is fixed for a particular frequency. Also, the zones can be chosen to yield two

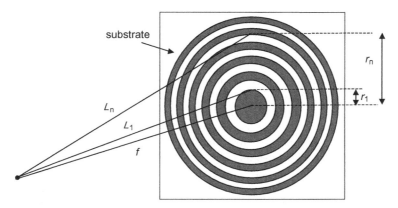

Figure 1.27. A Fresnel zone plate lens.

arbitrary foci, or where one focus is at infinity the zone plate is useful as a lens antenna. The lens radii can be derived from straightforward use of Pythagoras's theorem and Figure 1.27. Here, the path length L at each ring edge increments by one half wavelength, thus

$$L_n = f + n\frac{\lambda}{2}$$

where f is the focal distance and n is a counter = 1,2, . . .
and so:

$$r_n^2 = f^2 + \left(\frac{n\lambda}{2}\right)^2$$

Furthermore, for $f \gg \lambda$, such as might occur in optics, we can approximate $rn \approx \sqrt{nf\lambda}$.

The chief advantages of the zone plate, as a lens antenna, is the convenience of its flat structure and the simplicity of manufacture. Such a plate can easily be fabricated as a printed circuit, for example. Of course, its profound disadvantage is that the blockage of alternate zones rejects half the available power from the system. However, the concept has been adapted for microwave frequencies by using dielectric rings of alternating height where the adjacent zones' thicknesses differ by one half wavelength, and the metal (blocking) rings are dispensed with. This recovers the otherwise missing half of the available energy, but the phase relationship between rings (zones) is still imperfect, there being a 90° discontinuity at each boundary. These phase errors are reduced by reducing the granularity of the steps and leads to the phase correcting Fresnel zone plate lens Figure 1.18 discussed above, which typically uses more steps and more closely resembles the zoned dielectric lens.

In all these versions the complete zone-plate lens is made out of a single dielectric material. A recent alternative construction, based on the same principle, uses two different dielectric materials. In this version all of the rings have exactly the same thickness, but the phase-shifted rings are composed of the second dielectric material. The advantage of this structure is that the surface is completely flat instead of possessing small ridges. A recent article by J. C. Wiltse [58] discusses the operation of the zone-plate lens for millimeter waves and describes the various versions mentioned above. Another comprehensive summary is found in Reference 59.

Unsurprisingly, there exists a close analogue in reflector antennas with the Fresnel zone reflector which employs a similar zoning of reflective rather than diffractive bands.

The concepts of the Fresnel zone plate lens (Fig. 1.26) and the phase shifting surface lens were cleverly combined in Reference 60 where a phase shifting surface was used instead of the otherwise opaque metal zones, and a significant fraction of the lens area was not metalized at all. This structure was called a phase-correcting phase shifting surface (PSS) lens antenna and Gagnon et al. [60] compared it with three other lens types: dielectric plano-hyperbolic lens antenna, 90° phase-correcting Fresnel zone plate antenna. (Fig. 1.18) and Fresnel zone plate antenna (Fig. 1.27), all with a diameter of 152 mm and f/D ratio 0.5. Measurements on the 4 types were carried out in the region 30 GHz. The PSS lens of Gagnon et al. [60] outperformed all but the plano-hyperbolic dielectric lens antenna, falling short by just 0.3 dB of gain. However, it exhibited a weight reduction by a factor of almost ten and a forty-fold thickness reduction. The chief disadvantage was limited bandwidth, being 7% at 1 dB.

1.8 APPLICATIONS

Lens antennas can find application in a wide range of environments; a few are listed below:

1. Fixed wireless services such as telephony and internet. These services might be combined.
2. Satellite television (receive only service) with the possibility of multiple beams and hence using multiple satellites, without the need for motorized feeds.
3. Ultra Small and Very Small Aperture Terminals (USAT/VSAT) for two-way satellite communications.
4. In radar, for active or passive signal scanning of a wide area, for security purposes or for automotive collision avoidance radar such as is used in bands around 77 GHz.
5. Radio Frequency Identification (RFID).
6. Multi-beam antennas. In wireless local area network (WLAN) applications, multiple point-to-point communication links can be arranged using the same antenna, thus eliminating the slow-down of data speed when many nodes are transmitting in an omni-directional fashion and thus interfering with each other.

7. In compact radio transceivers where the electronic components and modem are protected in a casing behind the antenna aperture, offering a measure of isolation from the environment. Such an enclosed antenna is also excellent in repeater systems, where received and transmitted signals tend not to disturb each other.

Also, voltage standing wave ratio (VSWR) measurements reveal that the lens antenna is equally well-suited as a transmitter as a receiver. The reflection coefficient at the primary feed should be very low and so is not particularly likely to reflect much transmitted power back toward the power amplifier.

The compactness of lens antennas and the ease with which multi-beam variants can be produced makes them very attractive for exploiting the large bandwidths available at millimeter wave frequencies and thus suitable for very high data rate and spectrally efficient communications. This theme will occur several times throughout the following chapters.

1.9 ANTENNA MEASUREMENTS

Aside from an antenna's physical properties (mass, dimensions and so on) its electrical properties are of most interest to engineers and the measurement of which merits careful consideration. Also setting aside the property of return loss, which would entail the use of a vector network analyzer in a calibrated, single port measurement, it is the measurement of antenna radiation patterns which are of most interest. Approaches to measurement of radiation patterns fall into several broad areas such as indoor, outdoor, near field and far field methods. A summary may be found in Chapter 17 of Reference 17.

Outdoor ranges are, almost by definition, not protected from the environment (precipitation, interference), while indoor ranges offer limited inter-antenna separation. Indoor ranges use an anechoic interior lining to provide a non-reflective measurement environment . Two basic forms of anechoic chambers are rectangular and tapered box types; the latter are used more for low radio frequencies, for example, below 1 GHz.

1.9.1 Radiation Pattern Measurement

The radiation pattern of an antenna is three-dimensional over a sphere surrounding the antenna. Because it is not always practical to measure a three-dimensional pattern, a number of two-dimensional patterns (referred to as pattern cuts) are often measured. Different types of antenna positioners can perform this, such as Elevation-over-Azimuth, Azimuth-over-Elevation, or Roll-over-Azimuth mounts. Antenna directivity can be derived from the measured radiation pattern. This entails the numerical integration of the radiation intensity over all of space and so requires an adequate sampling over a spherical surface, which has implications for the time required to measure what can be a large data set. Alternatively, directivity might be inferred from two orthogonal planes of data, where interpolation is used for the space between and some loss of accuracy must be expected. For highly directive antennas (many lens antennas will fall in this category) measurement time may justifiably be reduced by capturing data over just the main lobe, again with some loss of accuracy.

1.9.2   Gain Measurement

Gain can be measured using two basic methods: either by comparison with a known calibration standard (e.g., a "standard horn"), or by calculation using the Friis transmission equation. The latter, also called the absolute gain method, requires that either the transmit and receive antennas are identical, or if they are different then three antennas and three measurements are needed to formulate a set of three simultaneous equations whose solution determines the gain of the antenna-under-test (AUT).

1.9.3   Polarization Measurement

A straightforward way to measure polarization entails two measurements using orthogonally polarized linear probes, which might be vertically and horizontally polarized (but need not be). In a second method, a single probe mechanically spins (perhaps quite quickly, e.g., a few Hz) which is quite useful for direct measurement of the axial ratio of a notionally circular polarized AUT.

1.9.4   Anechoic Chambers and Ranges

In choosing a measurement method one might best begin with an assessment of how conveniently a far field measurement could be performed. Here one would start with a calculation of the far field condition:

$$r_{far} > \frac{2D^2}{\lambda}$$

That is, the far field (*Fraunhofer*) zone of the antenna's radiation pattern can be approximately considered to begin at a distance not less than r_{far} above where D is the physical aperture diameter and λ is wavelength. More strictly, D is the diameter of a minimum radius sphere which wholly encloses the antenna structure. Furthermore, this approximation derives from the pragmatic observation that at this distance the maximum phase error between the observer and any part of the antenna aperture, of $\pi/8$ radians, would make little difference to the accuracy of measured patterns. Nevertheless, "little difference" could amount to 1 dB or more of error in measurement of sidelobe power at levels below about 25 dB with respect to peak main lobe gain. Often the right hand side of the above far field equation is modified to $\frac{4D^2}{\lambda}$ for improved results.

These simple formulae drive the required measurement distance for a far field measurement range. For distances of about 2–3 m anechoic chambers are quite practical and, relative to larger measurement ranges, quite low in cost. Many such small chambers are to be found in university research groups and commercial organizations. Often, either a single axis positioner (rotator) will be used to facilitate measurement of principle planes, or two-axis (e.g., elevation over azimuth positioners are used.)

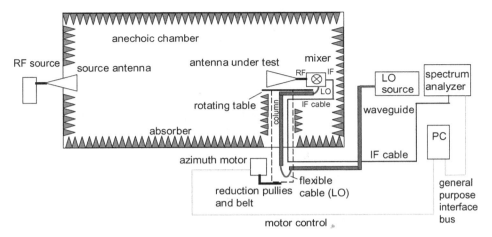

Figure 1.28. A single axis antenna pattern measurement system.

A relatively simple antenna measurement system is illustrated in Figure 1.28.

In Figure 1.28 the system is configured to measure power only—the spectrum analyzer is being used as a power meter. A motor drives a turntable in azimuth hence the AUT, the table to which it is mounted and the supporting column are rigidly connected. A computer (PC) commands the turntable motor in discrete angular increments and for each of these the received power is recorder, thus capturing a table of results with two columns: power vs. angle. The chamber is lined with absorbing material, typically a type of carbon-loaded foam of pyramidal shape (Fig. 1.29). The lengths of these pyramidal blocks influences the frequency range over which the absorber is effective. Shorter lengths (a few cm) with fine points are best for several GHz upwards, while longer lengths (a few 10s of cm) better for around 1 or 2 GHz. At lower frequencies the pyramidal points can be truncated if their shortest dimension is somewhat less than a wavelength. Any structures inside the chamber, for example, the turntable/ column housing, should be similarly screened such as the rotating column inside the tower structure in Figure 1.30.

Various approaches to signal routing can be chosen depending on available equipment and budget. In Figure 1.28 a system is illustrated where the power measurement is made at an intermediate frequency. This offers the advantage of very much reduced loss in the cable between the AUT and the power detector (spectrum analyzer). Of course, some other type of power meter could also be used. In contrast, a measurement of power at the RF frequency would tend to lead to a much more lossy circuit between AUT and detector, and require a more costly detector. On the other hand, a disadvantage of measuring at IF is the need to down convert the RF signal. In Figure 1.28 this entails use of a mixer and local oscillator (LO). A good cost compromise, shown here, is using a harmonic mixer which is driven at around half the RF frequency. To take an example, for an RF frequency of 30 GHz the LO could be around 14.5 GHz to produce an IF at

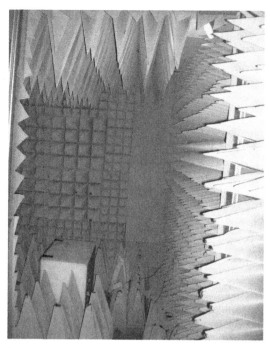

Figure 1.29. View of an anechoic chamber interior (photo: J. Thornton).

1 GHz. An important property of the mixer includes the required LO input power, which in turn influences the loss which can be tolerated in the LO circuit. If the mixer is a biased type (where direct current bias is applied to the mixer diodes) the LO power requirement is reduced. For non-biased mixers operating at the second harmonic, around + 10 dBm of LO power might be needed. The circuit between LO source and mixer might exhibit many dB of attenuation, but this can be minimized by substituting waveguide for cable as far as practically possible: this is indicated in Figure 1.28 where cable is only used for the rotating interface and the interconnections at the mixer and source. Another measure to ensure adequate LO power would be use of an amplifier (mixers can also be obtained with integrated LO amplifiers) so that loss in the LO circuit can be better tolerated.

All sorts of alternative configurations to that illustrated in Figure 1.28 would be practicable. The two sources (signal generators) could be replaced with a single item and power divided between RF and LO duties, implying of course that both are at the same frequency and so the mixer operates in the fundamental mode (not harmonic). Coherence of these two signals also implies a homodyne detection system and so one that is highly sensitive to the path length between source antenna and AUT and hence to any misalignment of the AUT phase center from the axis of rotation. Other laboratory configurations use a vector network analyzer effectively to perform both source and

Figure 1.30. Column in anechoic chamber screened by absorber (photo: J. Thornton.)

receiver functions and with the additional benefit of phase measurements should such be required.

Too small a chamber will prove restrictive for many electrically large antennas where longer distances are imposed by the far field condition. Some large organizations have constructed anechoic chambers on much larger scales, such as the European Space Agency's 12.5 × 8.5 × 4.5 m range at their European Space Research and Technology Centre (ESTEC) facility in The Netherlands (see photograph of Fig. 1.31). This chamber however is configured as a Compact Antenna Test Range (CATR), where far field conditions are brought about by use of a shaped reflector which yields a plane wave (or "quiet") zone in the proximity of the antenna under test. This reflector is seen on the right hand side of the image in Figure 1.31 and it is illuminated by an antenna on the balcony seen just left of center. The quiet zone where the positioner and AUT is placed is in the region on the very far left.

ESA's chamber is sufficiently large to contain entire satellites or payloads for analysis of payload antenna patterns, and for these purposes a multi-axis positioner is also installed. The CATR principle is also used in smaller chambers.

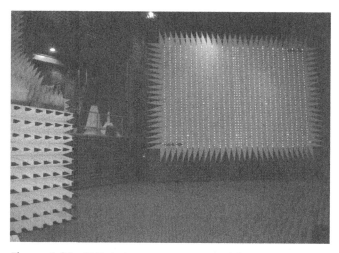

Figure 1.31. ESA's indoor compact range. (photo: J.Thornton).

For true far field measurement at longer ranges outdoor sites are used. Here, the principle of operation is conceptually the most straightforward, but the main issues are how to suppress or otherwise account for ground reflections, outside electromagnetic interference, and effects of the weather (either in terms of electrical attenuation and scattering, or environmental protection of equipment).

In contrast to far field or CATR methods, near field measurements are the other family of methods used for antenna pattern measurements and these are also well suited to indoor installation. Here the far field is not measured directly but derived from data gathered over a surface in the radiating near field (*Fresnel*) zone of the antenna under test (AUT). The surfaces most often encountered are spherical, cylindrical or planar. In each case the surface relates to both the mathematical transform used to derive far field results from near field data, and the mechanical properties of the scanning system. The data are derived as a set of two-port measurements where phase and amplitude are recorded for the transmission path between transmit and receive ports. In some cases the AUT is physically moved or rotated, or a probe in the measurement plane is moved, or combinations can occur.

Figure 1.32 illustrates the geometry of a typical spherical near field scanner. Here, either antenna may be the transmitter or receiver but most often the AUT is the receiver, as illustrated. The AUT is mounted on a mechanical two-axis rotator. One axis (e.g., θ) is first set, then the second (e.g., ϕ) swept through a circle and data recorded at fixed angular increments. The first angle is then incremented and the second swept through the full range. The order does not greatly matter, but determines whether data is captured around a set of circles resembling lines of latitude (Fig. 1.33a) or lines of longitude (Fig. 1.33b).

The equipment in the left hand and right hand sides in Figure 1.32 (i.e., the two antennas) may be located at opposite ends of a laboratory or chamber, or they

Figure 1.32. Spherical near field scan measurement.

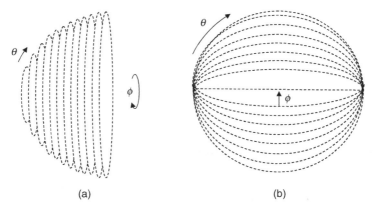

(a)   (b)

Figure 1.33. Spherical scans: (a) θ then φ, (b) φ then θ.

may be rigidly linked on a single piece of bench equipment such as that illustrated in Figure 1.34.

A variant of the near field class of methods is the "scalar only" near field type. Here, scalar data (amplitude only) is recorded over two surfaces at a fixed separation, for example over two planes which are a few wavelengths apart. While a single plane of scalar sampled near field data does not contain sufficient information to derive the far field, it is possible to infer the missing phase data where a second plane of scalar data is available. The process is iterative whereby the scalar data is successively transformed between planes, overwritten, and transformed again repeatedly so that the phase data "grows" in the matrix of samples. The method is considered analogous to holography. Convergence can be slow, but speeded up if other boundary conditions are invoked such as the truncation of the physical radiating aperture. The advantage is that

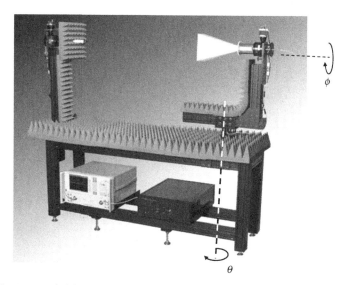

Figure 1.34. A near field spherical axis scanner (image courtesy of Nearfield Systems Inc, Torrance, CA, USA.)

the instruments used tend to be somewhat less costly since only a scalar detector is required.

Each of these three main categories also have different strengths and weaknesses, or areas of applicability. For example, the spherical method samples all the radiated field around an antenna and so can derive its far field pattern in all of space. As such, for mechanical reasons it is well suited for the measurement of smaller antennas, and for electrical reasons it is well suited to broad or quasi-omnidirectional patterns. Of course, these two properties (small/electrically small antenna and low directivity) tend to coincide. In contrast, the planar method samples data across a planar, usually rectangular zone close to the radiating aperture. As such, it can at best only derive far field pattern in one half of space, but in practice in an angular region somewhat less than this because of the finite extent of the spatial sampled area. It is well suited to electrically large and highly directional antennas. A very comprehensive treatise on planar near field measurements is found in Reference 61, while further details of far-field and near-field measurements in general can be found in Reference 62.

REFERENCES

[1] D. L. Sengupta and T. K. Sarkar, "Microwave and Millimeter Wave Research before 1900 and the Centenary of the Horn Antenna," Microwave Conference, 1995. 25th European, Vol. 2, pp. 903–909, 1995.

[2] F. G. Friedlander, "A dielectric-lens aerial for wide-angle beam scanning," IEE Electrical Engineers—Part IIIA: Radiolocation, pp. 658–662, 1946.

[3] K. Khoury and G. W. Heane, "High performance lens horn antennas for the millimeter bands," IEE Colloquium on Radiocommunications in the Range 30–60 GHz, 10/1–10/9, 1991.

[4] B. S. Westcott and F. Brickell, "General Dielectric-Lens Shaping Using Complex Co-ordinates," IEE Proceedings, Part H—Microwaves, Antennas and Propagation, Vol. 133, pt. H, no. 2, p. 122–126, April 1986.

[5] C. A. Fernandese, P. O. Frances, and A. M. Barbosa, "Shaped Coverage of Elongated Cells at Millimetrewaves Using a Dielectric Lens Antennas," Microwave Conference, 1995. 25th European, pp. 66–70, 1995.

[6] A. D. Olver and B. Philips, "Integrated Lens with Dielectric Horn Antenna," Electron. Lett. Vol. 39, 1993, pp. 1150–1152.

[7] N. G. Ugras, J. Zmuidzinas, and H. G. LeDuc, "Quasioptical SIS Mixer with a Silicon Lens for Submillimeter Astronomy," in Proceedings of 5th International Symposium Space Terahertz Technology, p. 125, 1994.

[8] J. W. Lamb, "Miscellaneous Data on Materials for Millimeter and Submillimeter Optics," Int. J. Inf. Millim. Waves Vol. 17, No. 12, 1996, pp. 1997–2034.

[9] C. R. Englert, M. Birk, and H. Maurer, "Antireflection Coated, Wedged, Single Crystal Silicon Aircraft Window for the Far-Infrared," IEEE Trans. Geosci. Remote Sens. Vol. 37, July 1999, pp. 1997–2003.

[10] A. J. Gatesman, J. Waldman, M. Ji, C. Musante, and S. Yngvesson, "An Anti-Reflection Coating for Silicon Optics at Terahertz Frequencies," IEEE Microw. Guid. Wave Lett. Vol. 10, No. 7, 2000, pp. 264–267.

[11] Dielectric Chart, Emerson & Cuming Microwave products, MA, 2010.

[12] H. Mosallaei and Y. Rahmat-Samii, "Nonuniform Luneburg and Two-Shell Lens Antennas: Radiation Characteristic Sand Design Optimization," IEEE Trans. Antennas Propag. Vol. 49, No. 1, January 2001, pp. 60–69.

[13] T. Komljenovic, R. Sauleau, and Z. Sipus, "Synthesizing Layered Dielectric Cylindrical Lens Antennas," APS Symposium, San Diego, July 2008.

[14] D. S. Weileand and E. Michielssen, "Genetic Algorithm Optimization Applied to Electromagnetics," IEEE Trans. Antennas Propag. Vol. 45, No. 3, March 1997, pp. 343–353.

[15] J. Robinson and Y. Rahmat-Samii, "Particle Swarm Optimization in Electromagnetics," IEEE Trans. Antennas Propag. Vol. 52, No. 2, February 2004, pp. 397–407.

[16] R. Sauleau and B. Barés, "A Complete Procedure for the Design and Optimization of Arbitrarily Shaped Integrated Lens Antennas," IEEE Trans. Antennas Propag. Vol. 54, No. 4, 2006, pp. 1122–1133.

[17] C. A. Balanis, Antenna Theory: Analysis and Design, 3rd ed., Wiley-Interscience, Hoboken, NJ, 2005.

[18] A. D. Olver, P. J. B. Clarricoats, A. A. Kishk, and L. Shafai, Microwave Horns and Feeds, IEEE Press, New York, 1994.

[19] C. Salema, C. Fernandes, and R. K. Jha, Solid Dielectric Horn Antennas, Artech House, Norwood, 1998.

[20] A. D. Olver and A. A. Saleeb, "Lens-Type Compact Antennas Range," Electron. Lett. Vol. 15, No. 14, 1979, pp. 409–410.

[21] S. Bishay, S. Cornbleet, and J. Hilton, "Lens Antennas with Amplitude Shaping or Sine Condition," Proc. IEE Vol. 136, No. 3, 1989, pp. 276–279.

[22] C. J. Sletten, Reflector and Lens Antennas—Analysis and Design Using Personal Computers, Artech House, Norwood, 1988.

[23] C. Fernandes, M. Filips, and L. Anunciada, "Lens Antennas for the SAMBA Mobile Terminal," Proceedings of the ACTS Mobile Telecommunications Summit 97, Aalborg, Denmark, pp. 563–568, 1997.

[24] R. K. Luneberg, United States Patent 2,328,157, Issue Date: August 31, 1943.

[25] M. Rayner, Use of Luneburg lens for low profile applications", Datron/Transco Inc. Microw. Product Dig., December 1999.

[26] X. Wu and J. J. Lauren, "Fan-Beam Millimeter-Wave Antenna Design Based on the Cylindrical Luneberg Lens," IEEE Trans. Antennas Propag. Vol. 55, No. 8, 2007, pp. 2147–2156.

[27] E. G. Zelkin and R. A. Petrova, Lens Antennas, Sovetskoe Radio Publ., Moscow, (in Russian), 1974.

[28] J. Fox, Proceedings of the Symposium on Quasi-Optics, Polytechnic Press, Brooklyn, NY, 1964.

[29] P. F. Goldsmith, "Quasioptical Techniques," Proc. IEEE Vol. 80, No. 11, 1992, pp. 1729–1747.

[30] J. C. G. Lesurf, Millimeter- Wave Optics, Devices, and Systems, Adam Hilger, Bristol, 1990.

[31] O. J. Lodge and J. L. Howard, "On Electric Radiation and its Concentration by Lenses," Proc. Physical Soc. Lond. Vol. 10, No. 1, 1888, pp. 143–163.

[32] E. Karplus, "Communication with quasioptical waves," Proc. IRE Vol. 19, October 1931, pp. 1715–1730.

[33] F. Sobel, F. L. Wentworth, and J. C. Wiltse, "Quasi-Optical Surface Waveguide and Other Components for the 100 to 300 Gc Region," IEEE Trans. Microw. Theory Tech. Vol. MIT-9, No. 6, 1961, pp. 512–518.

[34] E. Wolf, "Microwave Optics," Nature Vol. 172, 1953, pp. 615–616.

[35] H. Kosaka, T. Kawashlma, A. Tomita, M. Notami, T. Tomamura, T. Sato, and S. Kawakami, "Superprism Phenomena in Photonic Crystals: Toward Microscale Lightwave Circuits," J. Lightwuve Technol. Vol. 17, No. 11, 1999, pp. 2032–2038.

[36] J. B. Pendry, "Negative Refraction Index Makes a Perfect Lens," Phys. Rev. Lett. Vol. 85, 2000, pp. 3966–3969.

[37] J. Volakis, Antenna Engineering Handbook, 4th ed., McGraw-Hill Professional, New York, June 2007.

[38] A. Petosa and A. Ittipiboon, "Shadow Blockage Effects on the Aperture Efficiency of Dielectric Fresnel Lenses," IEE Proc.-Microw. Antennas Propag. Vol. 147, No. 6, December 2000, pp. 451–454.

[39] B.-I. Wu, W. Wang, J. Pacheco, X. Chen, T. Grzegorczyk, and J. A. Kong, "A Study of Using Metamaterials as Antenna Substrate to Enhance Gain," PIER Vol. 51, 295–328, 2005, pp. 295–329.

[40] D. R. Smith, W. J. Padilla, D. C. Vier, S. C. Nemat-Nasser, and S. Schultz, "Composite Medium with Simultaneously Negative Permeability and Permittivity," Phys. Rev. Lett. Vol. 84, No. 18, May 2000, pp. 4184–4187.

[41] D. Staiculescu, N. Bushyager, and M. Tentzeris, "Microwave/ Millimeter Wave Metamaterial Development Using the Design of Experiments Technique," IEEE Applied Computational Electromagnetics Conference, pp. 417–420, April 2005.

[42] E. Ozbay, K. Aydin, E. Cubukcu, and M. Bayindir, "Transmission and Reflection Properties of Composite Double Negative Metamaterials in Free Space," IEEE Trans. Antennas Propag. Vol. 51, No. 10, October 2003, pp. 2592–2595.

[43] R. A. Shelby, D. R. Smith, and S. Schultz, "Experimental Verification of a Negative Index of Refraction," Science Vol. 292, April 2001, pp. 77–79.

[44] G. V. Eleftheriades, A. K. Iyer, and P. C. Kremer, "Planar Negative Refractive Index Media Using Periodically L-C Loaded Transmission Lines," IEEE Trans. Microw. Theory Tech. Vol. 50, No. 12, December 2002, pp. 2702–2712.

[45] A. Grbic and G. V. Eleftheriades, "Experimental Verification of Backward-Wave Radiation from a Negative Refractive Index Material," J. Appl. Phys. Vol. 92, No. 10, November 2002, pp. 5930–5935.

[46] C. Caloz and T. Itoh, "Novel Microwave Devices and Structures Based on the Transmission Line Approach of Meta-Materials," IEEE MTT-S Digest, pp. 195–198, 2003.

[47] C. Metz, "Phased Array Metamaterial Antenna System," US Patent US2005225492 (A1) Issued on October 25, 2005.

[48] T. Ueda, A. Lai, and T. Itoh, "Negative Refraction in a Cut-off Parallel-Plate Waveguide Loaded with Two-Dimensional Lattice of Dielectric Resonators," 36th European Microwave Conference, pp.435–438, Manchester, U.K., September 2006.

[49] S. Kamada, N. Michishita, and Y. Yamada, "Metamaterial Lens Antenna Using Dielectric Resonators for Wide Angle Beam Scanning," IEEE Antennas and Propagation Society International Symposium (APSURSI), pp. 1–4, 2010.

[50] D. M. Pozar, "Flat Lens Antenna Concept Using Aperture Coupled Microstrip Patches," Electron. Lett. Vol. 32, No. 23, November 1996, pp. 2109–2111.

[51] M. R. Chaharmir, J. Shaker, M. Cuhaci, and A. Sebak, "Mechanically Controlled Reflectarray Antenna for Beam Switching and Beam Shaping in Millimeter Wave Applications," Electron. Lett. Vol. 39, No. 7, 2003, pp. 591–592.

[52] J. Huang and A. Feria, "A One-Meter X-band Inflatable Reflectarray Antenna," Microw. Opt. Technol. Lett. Vol. 20, January 1999, pp. 97–99.

[53] J. Huang and A. Feria, "Inflatable Microstrip Reflectarray Antennas at X and Ka-band Frequencies," IEEE AP-S, Orlando, Florida, pp. 1670–1673, July 1999.

[54] J. Thornton and D. J. Edwards, "Modulating Retroreflector as a Passive Radar Transponder," Electron. Lett. Vol. 34, No. 19, September 1998, pp. 1880–1884.

[55] B. A. Munk, Frequency Selective Surfaces: Theory and Design, John Wiley and Sons Inc., New York, 2000.

[56] N. Gagnon, A. Petosa, and D. A. McNamara, "Thin Microwave Quasi-Transparent Phase-Shifting Surface (PSS)," IEEE Trans. Antennas Propag. Vol. 58, No. 4, April 2010, pp. 1193–1201.

[57] A. Abbaspour-Tamijani, K. Saranbandi, and G. M. Rebeiz, "A Millimeter-Wave Bandpass Filter-Lens Array," IET Microw. Antennas Propag. Vol. 1, No. 2, April 2007, pp. 388–395.

[58] J. C. Wiltse, "Fresnel Zone-Plate Lenses," Proceedings of SPIE, Vol. 544, Millimeter-Wave Technology III, July 1985.

[59] H. D. Hristov, Fresnel Zones in Wireless Links, Zone Plate Lenses and Antennas, Artech House, Norwood, MA, 2000.

[60] N. Gagnon, A. Petosa, and D. A. McNamara, "Comparison between Conventional Lenses and an Electrically Thin Lens Made Using a Phase Shifting Surface (PSS) at Ka Band,"

Loughborough Antennas & Propagation Conference, pp.117–120, 16–17 November 2009, Loughborough, UK.

[61] S. Gregson, J. McCormick, and C. Parini, Principles of Planar Near-Field Antenna Measurements, IET Electromagnetic Wave Series 53, The Institute of Engineering and Technology, London, 2007.

[62] G. Evans, Antenna Measurement Techniques, Artech House, Norwood, MA, 1990.

2

REVIEW OF ELECTROMAGNETIC WAVES

Kao-Cheng Huang

Although this book is written for readers with a moderate knowledge of electromagnetic theory, it is useful, for consistency in notation and easy referencing, to dedicate a chapter to the fundamentals of electromagnetic wave theory and antenna parameters. Most of the material presented here may be found in classical textbooks such as Orfanidis's [1].

Starting with a review of Maxwell's equations in Section 2.1, we review the behavior of fields at boundaries and the equivalence theorem. A summary of antenna parameters (e.g., beam solid angle and antenna temperature) is presented in Section 2.2. Following the contents of polarization in Section 2.3, the chapter presents a very brief discussion of metamaterials in Section 2.4.

2.1 MAXWELL'S EQUATIONS

The history of antenna theory dates back to James Clerk Maxwell, who unified the theories of electricity and magnetism and eloquently represented their relations through a set of profound equations best known as *Maxwell's equations*. The work was first

Modern Lens Antennas for Communications Engineering, First Edition. John Thornton and Kao-Cheng Huang.
© 2013 Institute of Electrical and Electronics Engineers. Published 2013 by John Wiley & Sons, Inc.

published in 1873 [2], albeit not in the form of the modern and more concise vector notation used here. Maxwell's equations are nonetheless at the heart of all classical electromagnetic phenomena:

Gauss's law for electricity	$\nabla \cdot (\varepsilon E) = \rho$	Relates net electric flux to net enclosed electric charge
Gauss's law for magnetism	$\nabla \cdot (\mu H) = \rho_m = 0$	Relates net magnetic flux to net enclosed magnetic charge
Faraday's law	$\nabla \times E = -J_m - \mu \dfrac{\partial H}{\partial t}$	Relates induced electric field to changing magnetic flux
Ampere–Maxwell law	$\nabla \times H = +J + \varepsilon \dfrac{\partial E}{\partial t}$	Relates induced magnetic field to changing electric flux and to current

$$(2.1)$$

where

E is the electric field (volt per meter);

ρ is the electric charge density (coulomb per cubic meter);

ρ_m is the magnetic charge density (virtual);

H is the magnetic field (ampere per meter);

J is the electric current density (ampere per square meter);

J_m is the magnetic current density (virtual);

ε is permittivity, $\varepsilon = \varepsilon_r \varepsilon_0 (\varepsilon_r = 1$ in free space); and

μ is permeability, $\mu = \mu_r \mu_0$ ($\mu_r = 1$ in free space).

In Equation (2.1), electric current density J, magnetic current density M, electric charge density ρ, and magnetic charge density ρ_m are allowed to represent the sources of excitation. The respective current and charge densities are related by the *continuity equations*:

$$\nabla \cdot J = -\frac{\partial \rho}{\partial t} = -j\omega\rho$$

$$\nabla \cdot J_m = -\frac{\partial \rho_m}{\partial t} = -j\omega\rho_m,$$

$$(2.2)$$

where ω is the angular frequency.

Although magnetic point sources do not exist in the real world, they are often introduced as *electrical equivalents* to facilitate solutions of physical boundary-value problems. In fact, for some configurations, both electric and magnetic equivalent current densities are used to represent actual antenna systems. For metallic wire antennas, such as a dipole, an electric current density is used to represent the antenna. On the other hand, an aperture antenna, such as a lens or a horn, can be represented by either an equivalent magnetic current density or by an equivalent electric current density, or both. This satisfies Maxwell's equations in Equation (2.1).

When the charge density ρ and current density J work as the sources of the electromagnetic fields, these densities are constrained in space (e.g., they are restricted to flow on the antenna only). The generated electric and magnetic fields are radiated away from these sources and can propagate to the receiving antennas at large distances.

- Source-Free Fields

Away from the sources, that is, in source-free regions of space, Maxwell's equations are simplified as follows:

$$\nabla \times E = -\mu_0 \frac{\partial H}{\partial t} = -\frac{\partial B}{\partial t}$$

$$\nabla \times H = \varepsilon_0 \frac{\partial E}{\partial t} = \frac{\partial D}{\partial t} \qquad \text{(source-free Maxwell equations)}, \qquad (2.3)$$

$$\nabla \cdot D = 0$$

$$\nabla \cdot B = 0$$

where

D is the electric flux density ($D = \varepsilon_0 E$ in free space) and
B is the magnetic flux density ($B = \mu_0 H$ in free space).

The first two equations of Equation (2.3) enforce the boundary conditions on the discontinuity of the tangential components of the electric and magnetic fields, while the last two equations of Equation (2.3) imply the boundary conditions on the discontinuity of the normal components of the electric and magnetic flux densities.

- Time-Harmonic Fields

For time-harmonic fields, all four conditions of Equation (2.3) would relate to each other. The first two equations of Equation (2.3) form an independent and sufficient set as can be seen as follows:

$$\nabla \times E(r, t) = -\mu_0 \frac{\partial H(r, t)}{\partial t} \qquad (2.4)$$

$$\nabla \times H(r, t) = \varepsilon_0 \frac{\partial E(r, t)}{\partial t} + J(r, t). \qquad (2.5)$$

In addition to the boundary conditions of Equation (2.3), the solutions for the fields radiated by the antenna must also satisfy the radiation condition: The waves travel outwardly from the source in an infinite homogeneous medium and vanish at infinity.

If current density $J(r, t)$ is static, it is written as $J(r)$, which causes a spatially curling magnetic field $H(r)$; it fails to generate a temporally varying magnetic field, which means that $\partial H(r)/\partial t = 0$. According to Equation (2.4), this, in turn, fails to generate a spatially and temporally varying electric field $E(r)$. Therefore, a magnetic field is only generated in the location where we have a current density $J(r)$ present. Since we are only interested in making a wave propagating in a wireless environment where no charges (and hence current densities) can be supported, a static current density $J(r)$ is of little use.

On the other hand, if there is a time-varying current density $J(r, t)$, it generates a spatially and temporally varying magnetic field $H(r, t)$, according to Equation (2.5). Clearly, $\partial H(r, t)/\partial t \neq 0$, which, according to Equation (2.4), generates a spatially and temporally varying electric field, $E(r, t)$, that is, $\partial E(r, t)/\partial t \neq 0$. With reference to Equation (2.5), this phenomenon again generates a spatially and temporally varying magnetic field, $H(r, t)$, even in the absence of a current density, $J(r, t)$.

Hence, an electromagnetic wave is generated where the electric field stimulates the magnetic field and vice versa. It is obvious that such a wave can now propagate in space without the need of a charge-bearing medium; however, such a medium can certainly enhance or reduce the strength of the electromagnetic wave by influencing the current density $J(r, t)$. To make an electromagnetic wave decouple from a transmitting antenna, we need a medium capable of carrying a time-varying current density $J(r, t)$. A medium that achieves this with a high efficiency is called an *antenna*. An antenna can hence be a lens, a polyrod, metallic volumes and surfaces, and so on.

It is known that an electric current, $J(r, t)$, gives rise to a flow of electric charge whose rate is

$$J(r, t) = \partial Q/\partial t,$$

where Q is the electric charge.

If we want a time-varying current density for which $\partial J(r, t)/\partial t \neq 0$, we need to ensure that $\partial^2 Q/\partial t^2 \neq 0$. This means that we need to accelerate or decelerate charges. Such acceleration is achieved by means of a harmonic carrier frequency as deployed in conventional wireless communications systems or a noncarrier Gaussian pulse as deployed in ultrawide band systems. In both systems, the second-order differential exists and is different from zero, thereby facilitating a decoupling of the signal from the antenna.

Such charge acceleration can be brought about in bent wires where electrons following the outer radius move faster than those following the inner radius of the bend. Therefore, any nonstraight piece of wire will radiate electromagnetic waves, which is desirable in the case of antenna design.

Once the wave is launched into space, that is, decoupled from the antenna, it *propagates* through free space according to Equations (2.4) and (2.5) until it impinges upon and interacts with an object or reaches a receiver antenna.

2.1.1 Boundary Conditions

Most lens antennas are made of dielectric materials. The electric flux density D related to the electric polarization properties of the material:

$$D = \varepsilon_0 E_{1n} + P,$$

where the quantity P represents the dielectric polarization of the material, that is, the average electric dipole moment per unit volume.

The boundary conditions for the electromagnetic fields across material boundaries are given below:

$$
\begin{aligned}
E_{1t} - E_{2t} &= 0 \\
H_{1t} - H_{2t} &= J_s \times \widehat{n} \\
D_{1n} - D_{2n} &= \rho_s \\
B_{1n} - B_{2n} &= 0,
\end{aligned}
$$
(2.6)

where

- ρ_s is any external *surface charge* on the boundary surface (coulomb per square meter);
- J_s is any external *surface current densities* on the boundary surface (ampere per meter); and
- $\widehat{n}$ is a unit vector normal to the boundary pointing from the lower medium into the upper medium (Fig. 2.1).

In other words, the tangential components of the E-field are continuous across the interface; the difference of the tangential components of the H-field is equal to the surface current density; the difference of the normal components of the electric flux density D is equal to the surface charge density; and the normal components of the magnetic flux density B are continuous.

The boundary condition for the normal components of the electric flux density may also be written in a form that brings out the dependence on the *polarization surface charges*:

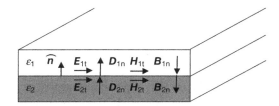

Figure 2.1. Boundary conditions between two dielectrics.

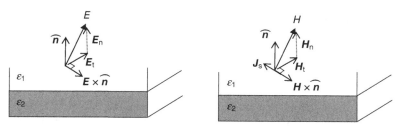

Figure 2.2. *E*-field (left) and *H*-field (right) directions at boundary.

$$D_{1n} - D_{2n} = \rho_s$$
$$\Rightarrow (\varepsilon_0 E_{1n} + P_{1n}) - (\varepsilon_0 E_{2n} + P_{2n}) = \rho_s \tag{2.7}$$
$$\Rightarrow \varepsilon_0 (E_{1n} - E_{2n}) = \rho_s - P_{1n} + P_{2n} \equiv \rho_{s,\text{sum}}.$$

The sum of surface charge density will be $\rho_{s,\text{sum}} = \rho_s + \rho_{1s,\text{pol}} + \rho_{2s,\text{pol}}$, where the surface charge density of polarization charges accumulating at the surface of a dielectric is

$$P_{s,\text{pol}} = P_n = \hat{n} \cdot P,$$

where $\hat{n}$ is the outward normal from the substrate dielectric.

The relative directions of the field vectors are shown in Figure 2.2. Each vector may be decomposed as the sum of a part tangential to the surface and a part perpendicular to it; that is,

$$E = E_t + E_n,$$

where

$$E_t = \hat{n} \times (E \times \hat{n}) \qquad \textit{tangential} \text{ components}$$
$$E_n = \hat{n}(\hat{n} \cdot E) = \hat{n} E_n \qquad \textit{normal} \text{ components.} \tag{2.8}$$

Using these results, we can write the first two boundary conditions of Equation (2.6) in the following vectorial forms, where J_s is purely tangential:

$$\begin{cases} \hat{n} \times (E_1 - E_2) = 0 \\ \hat{n} \times (H_1 - H_2) = J_s. \end{cases} \tag{2.9}$$

The boundary conditions (Eq. 2.6) can be derived from the integrated form of Maxwell's equations if we make some additional regularity assumptions about the fields at the interfaces. In many interface problems, there are no externally applied surface charges or currents on the boundary. In such source-free cases, the boundary conditions may be simplified as

$$E_{1t} = E_{2t}$$
$$H_{1t} = H_{2t}$$
$$D_{1n} = D_{2n}$$
$$B_{1n} = B_{2n}.$$

(2.10)

2.1.2 Equivalence Theorem

It is well-known that a given distribution of currents and charges originates a unique electromagnetic field. The inverse, however, is not true. Identical electromagnetic fields may be obtained by different source distributions. To solve some radiation problems, it may be easier to substitute the original field sources with others that are equivalent, that is, that produce the same radiation fields in a given region of space. The equivalence theorem is also known as the Schelkunoff equivalence principle.

Assume that the field sources are contained in a limited volume V_1 (e.g., a sphere) enclosed by surface S. The volume outside S will be denoted by V_2. The field sources (J, ρ) originate a unique electromagnetic field (E, H) both inside and outside V_1. From the boundary conditions (Eq. 2.10), we know that the tangential components of E and H are continuous on the surface S.

Consider sources that give rise to fields (E_1, H_1) inside V_1 and (E_2, H_2) inside V_2. For these sources to be equivalent to first ones, we must have everywhere in V_2

$$E_2 = E$$

$$H_2 = H.$$

The fields of the equivalent sources must satisfy Maxwell's equations. Any discontinuity of the tangential components of the magnetic field on the surface S must be compensated by an electric surface current density given by (Fig. 2.3)

$$J = n \times (H - H_1).$$

Similarly, any discontinuity of the tangential components of the electric fields on S must give rise to a magnetic surface current density:

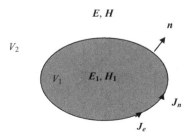

Figure 2.3. Equivalent sources.

$$J_m = -n \times (E - E_1).$$

Thus, an identical electromagnetic field may be produced by different source distributions. An investigation into the validity of the application of Schelkunoff's *equivalence principle* to a cylindrical dielectric surface appears in Reference 3. This article states that all treatments of a uniform cylindrical rod antenna that assume a radiation mechanism from the radial surface like those of a leaky waveguide are fallacious. Rather, the article states that radiation occurs only at discontinuities (and so the uniform rod cannot radiate).

2.2 ANTENNA PARAMETERS

2.2.1 Beam Solid Angle and Antenna Temperature

The power of electromagnetic waves emanating from an antenna is subject to the spatial direction; this phenomena is called the *radiation pattern*. For a lossless antenna, the *beam solid angle* Ω_A (Fig. 2.4) is defined as

$$\Omega_A = \int_{4\pi} P_n(\theta, \phi) d\omega = \frac{4\pi}{\text{Maximum gain}},$$

where $P_n(\theta, \phi)$ is the power pattern normalized to unity maximum.

So we see here the relationship that the higher the gain, the narrower the beam or power pattern.

A convenient practical unit for the power output per unit frequency from a receiving antenna is the *antenna temperature* (T_A). This term is commonly encountered in satellite communications engineering. Antenna temperature has nothing to do with the physical temperature of the antenna as measured by a thermometer; it is only the temperature of a matched resistor whose thermally generated power per unit

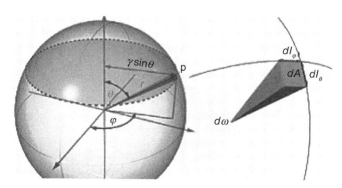

Figure 2.4. Beam solid angle.

frequency equals that produced by the antenna. The unit is widely used because of the following:

1. One Kelvin (K) of antenna temperature is a conveniently small power P_ν. $T_A = 1$ K corresponds to

$$P_\nu = k_B T_A = 1.38 \times 10^{-23} \text{ J/K} \times 1 \text{ K} = 1.38 \times 10^{-23} \text{ W/Hz}.$$

where k_B is *Boltzmann's constant* relating temperature and energy.

2. It can be calibrated by a direct comparison with different matching resistors connected to the receiver input.
3. The units of receiver noise are also Kelvin, so comparing the signal in Kelvin with the receiver noise in Kelvin makes it easy to assess whether a wanted signal can be detected above thermal noise.

Antenna temperature is defined as follows:

$$T_A \equiv \frac{P_\nu}{k_B}. \tag{2.11}$$

For an unpolarized point source, flux density S_f increases the antenna temperature by

$$T_A = \frac{P_\nu}{k_B} = \frac{A_E S_f}{2k_B}, \tag{2.12}$$

where A_E is the *effective collecting area*. The effective area A_E corresponding to a sensitivity of 1 K is

$$A_E = \frac{2k_B T}{S_f} = \frac{2 \times 1.38 \times 10^{-23} \text{ JK} \times 1 \text{ K}}{10^{-26} \text{ W/m}^2/\text{Hz}} = 2761 \text{ m}^2.$$

The beam solid angle is a useful parameter for estimating the antenna temperature, which is produced by a compact source covering solid angle Ω_B and having uniform brightness temperature T_B. The source should be much smaller than the beam so that the variation of P_ν is small across the source. The power per unit bandwidth received from the source by the receiving antenna is [4]

$$P_\nu \approx \frac{\lambda^2 G_{(max)} k_B T_B \Omega_B}{4\pi\lambda^2} = k_B T_B \frac{\Omega_B}{\Omega_A}.$$

Thus, the antenna temperature $T_A = P_\nu/k_B$ is

$$T_A \approx T_B \frac{\Omega_B}{\Omega_A}. \tag{2.13}$$

Equation (2.13) shows that the antenna temperature equals the source brightness temperature T_B multiplied by the fraction of the beam solid angle filled by the source. A source with $T_B = 104$ K covering 1% of the beam solid angle will add 100 K to the antenna temperature.

2.2.2 Directivity and Gain

In the antenna theory, an isotropic radiator is a theoretical radiator having a directivity of unity, or 0 dBi (decibels relative to an isotrope), which means that the radiator equally transmits (or receives) electromagnetic radiation to or from all directions. The directivity of an antenna is given by the ratio of the maximum radiation intensity (power per unit solid angle) to the average radiation intensity (averaged over a sphere). The directivity of any source, other than an isotrope, is always greater than unity in some direction while less than unity in others.

In reality, the isotropic radiator, with a radiation pattern (as expressed in spherical coordinates) is

$$E(r, \theta, \phi) = \frac{e^{-jkr}}{4\pi r} \hat{u}(\theta, \phi). \tag{2.14}$$

In Equation (2.14), the magnitude is independent of the spherical angles θ and ϕ, but it is permissible for the vector's direction, as represented by the unit vector $\hat{u}$ to be a function of θ and ϕ.

For a wireless network, it is generally assumed that the signal arriving at the receiver consists of many copies of the information-carrying signal, which have been generated by scattering and other processes by the environment. Each path will have a specific delay, and arrival times will vary according to the dimensions of the environment.

If a multipath fading effect should be avoided (e.g., in millimeter wave communications), it is necessary to focus or direct the radiated power from antennas in a given direction. The power flux density in this direction will be greater than if it were an omnidirectional antenna transmitting the same power (the power presented at the antenna input terminals) and the ratio between these values (i.e., the degree to which the antennas enhance the power flux density relative to an isotropic radiator) is called the antenna gain [4]. The maximum gain of an antenna is simply defined as the product of the directivity and its radiation efficiency. It can be expressed as follows:

$$\text{Gain} = \text{directivity} \times \text{efficiency.}$$

If the efficiency is not 100%, the gain is less than the directivity. A typical antenna is able to couple energy to and from free space with an efficiency of approximately 65%.

The usual approach to establish the power received by an antenna is to consider an isotropic radiator transmitting power P_T so that this power is distributed over the surface of an expanding sphere as the wave propagates. At the receiver, the power flux density (power per unit area) is then $P_T/4\pi R^2$. The received power is then determined

TABLE 2.1. Comparison of Different Types of Antennas

	Power Gain	Polarization
Lens antenna	High	Linear/circular
Polyrod antenna	High	Linear/circular
Printed antenna	Medium	Linear/circular
Horn antenna	High	Linear
Helical antenna	Medium	Circular
Multidipole	Medium	Linear/circular
Dipole	Low	Linear
Slot antenna	Low	Linear/circular

TABLE 2.2. Two- and Three-Dimensional Feeding Methods to Increase Antenna Gain

	Polyrod	Hemisphere Lens
2-D feed (e.g., patch)	Reference 5	Figure 2.5, Reference 6.
3-D feed (e.g., horn)	Reference 7	Figure 1.11

by the *effective capture area* of the receive antenna. This effective capture area can, in turn, be related to the gain of the antenna [4], so the gain can be written as

$$G = \frac{4\pi A_{\mathrm{E}}}{\lambda^2}.$$

To aid in designing the appropriate antenna for the application, Table 2.1 compares lens antennas with other types of antenna in terms of power gain and polarization.

For an antenna with a 20° half-power beamwidth (symmetric), the directivity can be calculated approximately to be

$$D = \frac{40,000}{\theta_{\mathrm{HPBW}}\phi_{\mathrm{HPBW}}} = \frac{40,000}{20 \times 20} = 100 \text{ or } 20 \text{ dBi.} \tag{2.15}$$

As discussed in Chapter 1, lens antennas can be fed by horn (three-dimensional feed) or by microstrip (two-dimensional feed) antennas. In general, three-dimensional feeds provide higher gain than two-dimensional feeds. Typical designs are summarized in Table 2.2.

The lens in Figure 2.5 is made out of low-cost low-permittivity Rexolite material. The single-beam lens achieves a gain of 24 dBi at 30 GHz and a front-to-back ratio of 30 dB. An axial ratio of 0.5 dB is maintained within the main lobe [8]. The measured impedance bandwidth is 12.5% within a standing wave ratio (*SWR*) of 1.8:1.0. The single-beam antenna is well suited for broadband wireless point-to-point links. The lens, which is fed by multiple slots, can radiate multiple beams with a minimum 3-dB overlapping level among adjacent beams. The coverage of the lens antenna system has

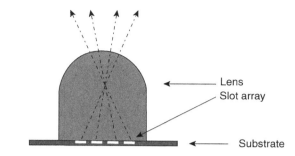

Figure 2.5. Multiple-beam launching through a substrate lens antenna.

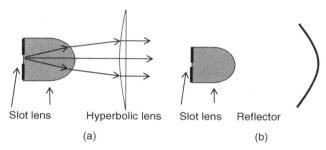

Figure 2.6. Schematic diagram of the two-antenna systems: (a) two-lens system, (b) lens-fed reflector.

been optimized through the utilization of a number of slot arrangements, leading to broad scan coverage. The multiple-beam lens antenna is suitable for an indoor wireless access point or as a switched beam smart antenna in portable devices.

When implementing a single-element antenna (such as a dielectric lens or a slot), it is possible to use two-antenna systems to achieve up to a 75% reduction of the lens material while maintaining about the same length and on-axis characteristics, as shown in Figure 2.6 [9]. The lens-fed reflector provides higher overall efficiency than the two-lens system. This makes the lens-fed reflector attractive for single-beam applications. In these two-antenna systems, a limited scan capability with multiple beams cross coupled at the 3-dB level is possible, which can lead to lower alignment requirements between a receiver and a transmitter for line-of-sight (LOS) broadband wireless links.

2.2.3   Antenna Beamwidth

The radiation pattern of an antenna is essentially the Fourier transform (linear space to the angle) of its aperture illumination function. There are four types of antenna configurations for a transmitter (Tx) and a receiver (Rx) in a communications system:

1. Tx (omnidirectional antenna) versus Rx (omnidirectional antenna)
2. Tx (omnidirectional antenna) versus Rx (directional antenna)

3. *Tx* (directional antenna) versus *Rx* (omnidirectional antenna)
4. *Tx* (directional antenna) versus *Rx* (directional antenna)

Omnidirectional antennas have signals radiating in all directions and are useful when a very wide field of view is needed for communications purposes, for example, with a mobile or handheld terminal where there is no control over orientation. In contrast, lens antennas are most often used as directional antennas which have a narrow beam in a desired direction and so reject signals from undesired directions. This is useful when wide coverage (or multipath) is not required. As a directional antenna has this small-angle coverage, it may, in some cases, be necessary to incorporate it with a beam-steering function to provide wider coverage or "tracking" if the terminal is moving. However, the narrower the antenna beamwidth, the more precise and hence complex the beam-tracking function would need to be. Hence, it is necessary to consider the balance between the complexities of a beamforming antenna and a tracking function.

In a radiation pattern cut containing the direction of the maximum of a lobe, the angle between the two directions in which the radiation intensity is one-half the maximum value (see Fig. 2.7) is called the half-power beamwidth. This is also commonly referred to as the 3-dB beamwidth. Beamwidth typically decreases as antenna gain increases.

2.2.3.1 Aperture Illumination and Beamwidth For a circular aperture antenna of diameter D, if the antenna is uniformly illuminated, the half-power beamwidth introduced above, in degrees, is about $55 \times$ wavelength/D; that is, beamwidth is directly proportional to the ratio of wavelength to diameter. This "uniform aperture" is often encountered as a theoretical concept but rather less often in practice. Among its various properties are the maximum gain and smallest beamwidth that can be achieved from a given aperture area, sometimes also called a "diffraction-limited" aperture. A more realistic aperture is one with a "tapered" illumination function, where

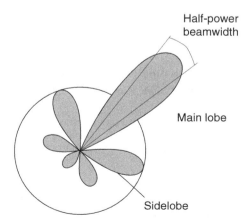

Figure 2.7. Beamwidth in a normalized power pattern (the radial scale is logarithmic).

the edge of the aperture is less strongly illuminated than its center. This typically arises from the illumination pattern of a feed incident upon a secondary aperture, for example, a reflector or lens. This taper leads to an increase in the beamwidth and suppressed sidelobes. Another approximation often encountered—it is really just a rule of thumb— is beamwidth (in degrees) = 70 × wavelength/D.

The exact beamwidth actually depends on the aperture illumination function. The next lobe in the pattern, usually called the first sidelobe, will be at about 17 dB below that of the main lobe for the uniform circular aperture, and a lower figure for the tapered aperture. Any further out sidelobes will usually have lower values and so tend to decay with increasing angle. The rate of decay of these sidelobes is an important parameter in many antenna applications and is used in many international standards as a defining parameter of antenna performance.

In general, the maximum gain can be approximated by the following formula:

$$G = \frac{27,000}{BW_h \times BW_v},$$ (2.16)

where G is the power gain (linear) of the antenna, BW_h is the horizontal beamwidth of the antenna, and BW_v is the vertical beamwidth of the antenna [4].

As an example, consider an antenna that has a vertical beamwidth of 27° and a horizontal beamwidth of 10°; it will have a power gain of 100 or 20 dB. It would also have a vertical dimension of about two wavelengths and a horizontal dimension of about five wavelengths if the antenna is uniformly illuminated.

2.2.3.2 *Free-Space Propagation*
The total received signal is normally expressed as a closed-form expression, known as the Friis equation [4]:

$$P_{Rx} = P_{Tx}\frac{G_{Tx}G_{Rx}\lambda^2}{16\pi^2 d^2 L},$$ (2.17)

where

P_{Rx} is the received power;

P_{Tx} is the transmitted power;

G_{Tx} and G_{Rx} are the antenna transmit gain and receive gain, respectively; and

λ, d, and L are the wavelength, separation, and other losses, respectively.

The allocations given to each of these components constitute what is generally called the *link budget*.

2.2.4 Aperture of a Lens

The aperture of an antenna is the area that captures energy from a passing radio wave. For a lens antenna, it is not surprising that the aperture is the size of the lens. For lens antennas that have apertures which are very large in terms of wavelength, it is frequently desirable to use a continuous type of aperture distribution because of the relative simplicity, as compared with a discrete-element type of array, which requires a large number of driven elements.

Power loss due to dielectric heating can be calculated by using the loss tangent and complex permittivity for the particular substrate dielectric material. For the half-wave dipole (a series-type resonance), for example, the radiation efficiency based on dielectric loss can be calculated as

$$\eta = \frac{R_r}{R_r + R_l},$$

where R_r is the radiation resistance at the input terminals and R_l is the loss resistance.

R_r and R_l can be found from the two calculations of the *input impedance*: one with $\tan \delta = 0$ and one with $\tan \delta \neq 0$. The radiation resistance is $R = \mathrm{Re}\,(Z_{in})$ for $\tan \delta = 0$, and the loss resistance is found from $R_r + R_l = \mathrm{Re}\,(Z_{in})$ with $\tan \delta \neq 0$. This is an accurate procedure for small losses. Note that the power loss due to surface waves is not considered in this case.

2.2.5 Phase Center

An antenna phase center is defined as a point such that, when the radiation patterns obtained when the antenna is rotated about it, exhibits constant phase at least for a considerable part of the main lobe (Fig. 2.8). For circular plane apertures with no phase error, the phase center lies on the aperture center. If the aperture exhibits a quadratic phase error, such as is the case of horns, the phase center is usually located slightly behind the aperture.

The position of the phase center may be found experimentally by trial and error, numerical approaches, or often nowadays using so-called solver software packages. For horns, the phase center position is normally located close to the aperture plane. The phase center of integrated lenses is very dependent upon the lens shape and may even be far out from the lens body. However, when bolometers or other detectors are used at the lens under test, only amplitude measurements can be obtained, preventing the usual phase center calculation from radiation pattern phase information [10].

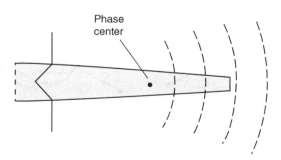

Figure 2.8. Phase center of a polyrod antenna.

2.3 POLARIZATION

The subject of antenna polarization has generated much published material over the years. The precise definition can be complex as radiating and receiving structures respond varyingly, both in frequency and the angle to incident and transmitted waves. Here, the discussion shall be confined to a simple treatment, and the reader is directed to texts that deal with the topic in much greater depth [11]. At millimeter wave and submillimeter wave range, only "far-field" radiation will be considered (since the wavelength is small compared with the dimensions of the radiators), and for the illustrative cases presented here, plane wave propagation will also be assumed.

In free space, the energy radiated by any antenna is carried by a transverse electromagnetic wave, which is composed of an electric field and a magnetic field. These fields are orthogonal to each another and also orthogonal to the direction of propagation. The electric field of the electromagnetic wave is used to define the polarization plane of the wave and therefore describes the polarization state of the antenna.

Consider a forward-moving wave and let $\boldsymbol{E}_0 = \hat{\boldsymbol{x}}A_+ + \hat{\boldsymbol{y}}B_+$ be its complex-valued phasor amplitude (A_+ and B_+ are complex numbers), so that $\boldsymbol{E}(z) = \boldsymbol{E}_0 e^{-jkz} = (\hat{\boldsymbol{x}}A_+ + \hat{\boldsymbol{y}}B_+)e^{-jkz}$. The time-varying field is obtained by restoring the factor $e^{j\omega t}$:

$$\boldsymbol{E}(z,t) = (\hat{\boldsymbol{x}}A_+ + \hat{\boldsymbol{y}}B_+)e^{jwt-jkz}. \tag{2.18}$$

The polarization of a plane wave is defined to be the direction of the electric field. For example, if $B_+ = 0$, the E-field is along the x-direction and the wave will be linearly polarized.

In describing antenna polarization properties, the *Ludwig definition 3* is commonly used [12]. In this definition, reference and cross polarizations are defined as the measurement obtained when antenna patterns are taken in the usual manner, as illustrated in Figure 2.9.

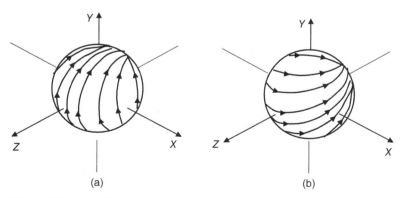

Figure 2.9. (a) Direction of the reference polarization. (b) Direction of the cross polarization (Ludwig 3).

This wave is said to be linearly polarized; that is, the electric field vector is confined to a single plane. Two independent linearly polarized waves at the same frequency can therefore exist and propagate along the same path. This feature has been used for many decades in free-space links, which utilize frequency reuse in order to double the capacity of a link for a given bandwidth. In this case, each polarization carries different information and is transmitted and received independently. Where the relative angular orientation of the transmitter and receiver is not defined, using two linear polarizations becomes a problem as the alignment of the receiver with the transmitter is essential. Systems have been deployed in which dynamic control of the receiver is used with the incident linear polarization, but coverage of these topics is beyond the scope of this book.

When the two polarizations carry the same information and the two components possess a specific phase relationship with each other, a form of wave can be constructed in which the electric field vector rotates as the wave propagates. If the relative phase of the two components is fixed at $\pm 90°$ and the amplitudes of the components are equal, the electric vector describes a circle as the wave propagates. Such a wave is said to be circularly polarized. The sense (or handedness) of the circular polarization depends on the sense of the phase shift. In general, the two linear components of the propagating wave can have an arbitrary (though constant) phase relationship and also different amplitudes. Such waves are said to be elliptically polarized.

The majority of electromagnetic waves in real systems are elliptically polarized. In this case, the total electric field of the wave can be decomposed into two linear components, which are orthogonal to each other, and each of these components has a different magnitude and phase. At any fixed point along the direction of propagation, the electric field vector will trace out an ellipse as a function of time. This concept is shown in Figure 2.10, where, at any instant in time, E_x is the component of the electric field in the x-direction and E_y is the component of the electric field in the y-direction. The total electric field E is the vector sum of Ex plus E_y. The projection along the line of propagation is shown in Figure 2.11.

Therefore, from the above discussion, there are two special cases of elliptical polarization, which are linear polarization and circular polarization. The term used to describe the relationship between the magnitudes of the two linearly polarized electric field components in a circularly polarized wave is the axial ratio (AR). In a pure circularly polarized wave, both electric field components have equal magnitude and the AR is 1 or 0 dB (10 log [AR]). Thus, in a pure linearly polarized wave, the axial ratio is ∞. In this case, the polarization ellipse traced by the wave is a circle.

It is difficult to make low cross-polarization circular sources that operate over a large bandwidth. Thus, cross polarization in the transmit antenna can be a major source of error in antenna gain. To deliver maximum power between a transmitter and a receiver antenna, both antennas must have the same angular orientation, the same polarization sense, and the same axial ratio. When the antennas are not aligned or do not have the same polarization, there will be a reduction in energy or power transfer between the two antennas. This reduction in power transfer will reduce the overall signal level, system efficiency, and performance. The polarization loss can affect the link budget in a communications system.

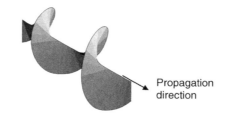

Figure 2.10. Propagation of elliptical polarization.

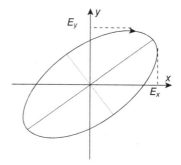

Figure 2.11. The projection of a polarization ellipse.

It is reported that circular polarization can reduce the power of the reflected path significantly in the millimeter wave LOS link [12]. Owing to the boundary conditions on the electric field, the in-plane and normal components of the electric field suffer a differential phase shift of 180° on reflection. This causes the sense of the circular polarization to be changed at each surface reflection. Thus, for an odd number of reflections, the reflected wave attains a polarization state orthogonal to the incident wave. When this occurs, a left-hand circularly polarized wave would become a right-hand circularly polarized wave and vice versa. However, the direction of circular polarization remains the same when there is an even number of reflections, and the power is only reduced due to reflection loss.

The circular polarization maintains an advantage in some user scenarios. For example, if a user holds the terminal at an arbitrary tilt angle to the transmit signal, there would be a degradation of the signal strength in the case of linear polarization signals. However, such degradation is not present in the case of circular polarization for a direct LOS and arbitrary terminal tilt angle. In addition, the terminal will receive fewer multipaths (for a single polarization), and circular polarization also offers the possibility of frequency reuse, albeit with the complication of cross-polar interference due to multipath reflections. Clearly, the magnitude of the multipaths depends on the reflection coefficients of the materials in the environment and on the material properties of the reflecting objects.

Conventional short-range systems normally use linearly polarized antennas to reduce cost. When the transmit and receive antennas are both linearly polarized, the physical antenna misalignment will result in a polarization mismatch loss, which can be determined using

$$\text{Polarization mismatch loss (dB)} = 10\log(\cos\theta),$$

where θ is the angular misalignment or tilt angle between the two antennas. Polarization efficiency can be written as

$$\text{Polarization efficiency} = 20\log\left(1\pm\frac{A_{r1}-1}{A_{r1}+1}\times\frac{A_{r2}-1}{A_{r2}+1}\right), \tag{2.19}$$

where A_r is the axial ratio and the subscripts the antenna number. Figure 2.12 illustrates some typical mismatch loss values for various misalignment angles.

In the circumstance where the transmitting antenna in a wireless link is circularly polarized and the receiving antenna is linearly polarized, it is generally assumed that a 3-dB system loss will result because of the polarization difference between the two antennas. In reality, the polarization mismatch loss between these two antennas will only be 3 dB when the circularly polarized antenna has an axial ratio of 0 dB. The actual mismatch loss between a circularly polarized antenna and a linearly polarized antenna will vary depending upon the axial ratio of the (nominally) circularly polarized antenna.

When the axial ratio of the circularly polarized antenna is greater than 0 dB (i.e., it is, in fact, elliptically polarized), this will dictate that one of the two linearly

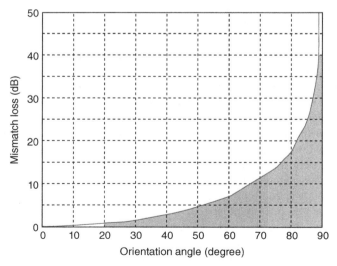

Figure 2.12. Polarization mismatch between two linearly polarized waves as a function of angular orientation θ.

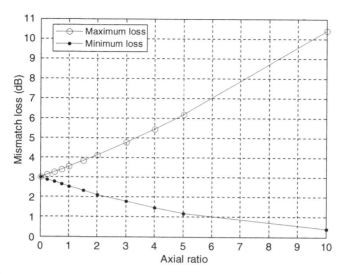

Figure 2.13. Polarization mismatch between a linearly and a circularly polarized wave as a function of the circularly polarized wave's axial ratio.

polarized axes will generate a linearly polarized signal more effectively than the other component. When a linearly polarized receiver is aligned with the polarization ellipse's major axis, the polarization mismatch loss will be less than 3 dB. When a linearly polarized wave is aligned with the polarization ellipse's linear minor axis, the polarization mismatch loss will be greater than 3 dB. Figure 2.13 illustrates the minimum and maximum polarization mismatch loss potential between an elliptically polarized antenna and a linearly polarized antenna as a function of the axial ratio. Minimum polarization loss occurs when the major axis of the polarization ellipse of the transmitter (receiver) is aligned with the plane of the linearly polarized wave of the receiver (transmitter). Maximum polarization loss occurs when the weakest linear field component of the circularly polarized wave is aligned with the linearly polarized wave.

An additional issue to consider with circularly polarized antennas is that their axial ratio will vary with the observation angle . Most manufacturers specify the axial ratio at the antenna boresight or as a maximum value over a particular range of angles. This range is generally chosen to represent the main beam of the antenna. In order to measure the axial ratio, the manufacturer would measure the antenna radiation pattern with a spinning linearly polarized source. As the source antenna spins, the difference in amplitude between the two linearly polarized wave components radiated or received by the antenna is evident. The resulting radiation pattern will describe the antenna's axial ratio characteristics for all measured observation angles.

A typical axial ratio pattern for a circularly polarized antenna is presented in Figure 2.14. From the antenna radiation pattern, it can be seen that the axial ratio at the boresight is about 0.9, while at an angle of +60° off-boresight, it dips to about 0.04. As the

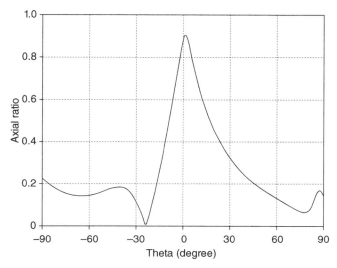

Figure 2.14. Typical axial ratio pattern for a helix antenna.

axial ratio varies with the observation angle, the polarization mismatch loss between a circularly polarized antenna and a linearly polarized antenna will vary with the observation angle as well.

However, in most cases, the polarization mismatch loss issue is rather more complex. Based upon this discussion of polarization mismatch loss, it would be conceivable that communication in a wireless system is near impossible when, for instance, the polarization of antenna of a mobile device is orthogonal to that of the access point (base station).

Obviously, this is unlikely to happen in the real world. In any mobile handset communications link, the signal between the handset antenna and the base station antenna is generally composed of a direct LOS signal and a number of multipath signals. In many instances, the LOS signal is not present and the entire communications link is established with multipath signals.

Multipath signals arrive at the antennas of mobile devices via the reflection of the direct signal off nearby and distant objects or surfaces. If the reflecting objects are oriented such that they are not aligned with the polarization of the incident wave, the reflected wave will experience a polarization state change. The resultant, or total signal, available to the receiver at the end of the communications link will be the vector summation of the direct signals and all of the multipath signals. In many instances, there will be a number of signals arriving at the receive site that are not aligned with the assumed standard polarization of the system antenna. Put another way, the scattering from the environment gives rise to random polarization. As the receive antenna rotates from vertical to horizontal, it simply intercepts or receives power from these multiple signals and will, in fact, receive different multipaths as the angle of orientation varies.

More precisely, polarization is the direction of the time-varying real-valued field. At any fixed point z, the vector $E(z, t)$ may be along a fixed linear direction or it may be rotating as a function of t, tracing a circle or an ellipse.

The polarization properties of the plane wave are determined by the relative magnitudes and phases of the complex-valued constants A_+, B_+ (see Eq. 2.18). Writing them in their polar forms $A_+ = Ae^{j\varphi_a}$ and $B_+ = Be^{j\varphi_b}$, where A and B are positive magnitudes, we obtain

$$E(z, t) = \left(\hat{x}Ae^{j\phi_a} + \hat{y}Be^{j\phi_b}\right)e^{jwt-jkz} = \hat{x}Ae^{j(wt-kz+\phi_a)} + \hat{y}Be^{j(wt-kz+\phi_b)}. \quad (2.20)$$

Extracting real parts and setting $E_R(z, t) = \mathrm{Re}\left[E(z, t)\right] = \hat{x}E_{Rx}(z, t) + \hat{y}E_{Ry}(z, t)$, we find the corresponding real-valued x, y-components:

$$\begin{aligned} E_{Rx}(z, t) &= A\cos(\omega t - kz + \phi_a) \\ E_{Ry}(z, t) &= B\cos(\omega t - kz + \phi_b). \end{aligned} \quad (2.21)$$

For a backward moving field, we replace k by $-k$ in the same expression. To determine the polarization of the wave, we consider the time dependence of these fields at some fixed point along the z-axis, say, at $z = 0$:

$$\begin{aligned} E_{Rx}(t) &= A\cos(\omega t + \phi_a) \\ E_{Ry}(t) &= B\cos(\omega t + \phi_b). \end{aligned} \quad (2.22)$$

The electric field vector $E_R(t) = \hat{x}E_{Rx}(t) + \hat{y}E_{Ry}(t)$ will be rotating on the xy-plane with angular frequency ω, with its tip tracing, in general, an ellipse. To see this, we expand Equation (2.22) using a trigonometric identity:

$$\begin{aligned} E_{Rx}(t) &= A[\cos \omega t \cos \varphi_a - \sin \omega t \sin \varphi_a] \\ E_{Ry}(t) &= B[\cos \omega t \cos \varphi_b - \sin \omega t \sin \varphi_b]. \end{aligned}$$

Solving for $\cos \omega t$ and $\sin \omega t$ in terms of $E_{Rx}(t)$, $E_{Ry}(t)$, we find

$$\begin{aligned} \cos \omega t \sin \phi &= \frac{E_{Ry}(t)}{B}\sin \phi_a - \frac{E_{Rx}(t)}{A}\sin \phi_b \\ \sin \omega t \sin \phi &= \frac{E_{Ry}(t)}{B}\cos \phi_a - \frac{E_{Rx}(t)}{A}\cos \phi_b, \end{aligned} \quad (2.23)$$

where we defined the relative phase angle $\varphi = \varphi_a - \varphi_b$.

Forming the sum of the squares of the two equations and using the trigonometric identity $\sin^2 \omega t + \cos^2 \omega t = 1$, we obtain a quadratic equation for the components E_x and E_y, which describes an ellipse on the E_{Rx}, E_{Ry} plane:

$$\left(\frac{E_{Ry}(t)}{B}\sin \phi_a - \frac{E_{Rx}(t)}{A}\sin \phi_b\right)^2 + \left(\frac{E_{Ry}(t)}{B}\cos \phi_a - \frac{E_{Rx}(t)}{A}\cos \phi_b\right)^2 = \sin^2 \phi.$$

This simplifies into

$$\frac{E_{Rx}^2}{A^2} + \frac{E_{Ry}^2}{B^2} - 2\cos\phi \frac{E_{Rx}E_{Ry}}{AB} = \sin^2\phi \text{ (polarization ellipse).} \qquad (2.24)$$

Subject to the values of the three quantities (A, B, φ), this polarization ellipse may be an ellipse, a circle, or a straight line. The electric field is accordingly called *elliptically*, *circularly*, or *linearly* polarized.

• Linear Polarization

In the case of linear polarization, we simply set $\varphi = 0$ or $\varphi = \pi$, corresponding to $\varphi_a = \varphi_b = 0$, or $\varphi_a = 0$, $\varphi_b = -\pi$, so that the phasor amplitudes are

$$\boldsymbol{E}_0 = \hat{\boldsymbol{x}}A \pm \hat{\boldsymbol{y}}B$$

and

$$E_{Ry} = \pm \frac{B}{A} E_{Rx}. \qquad (2.25)$$

• Circular Polarization

In the case of circular polarization, we set $A = B$ and $\varphi = \pm\pi/2$. Thus, the polarization ellipse (Eq. 2.25) becomes the equation of a circle:

$$\frac{E_{Rx}^2}{A^2} + \frac{E_{Ry}^2}{A^2} = 1. \qquad (2.26)$$

2.4 WAVE PROPAGATION IN METAMATERIALS

The permittivity ε and permeability μ of metamaterials are simultaneously negative, $\varepsilon < 0$ and $\mu < 0$ (see Section 1.6). Veselago was the first to study their unusual electromagnetic properties, such as having a negative index of refraction and the reversal of Snell's law [13–15].

The novel properties of metamaterials and their potential applications have attracted a lot of research interest [15]. Examples have been constructed using periodic arrays of wires and split-ring resonators [16] and by transmission line elements [17], and have been shown to exhibit the properties predicted. When $\varepsilon_{rel} < 0$ and $\mu_{rel} < 0$, the refractive index, $n^2 = \varepsilon_{rel}\mu_{rel}$, must be defined by the negative square root

$$n = -\sqrt{\varepsilon_{rel}\mu_{rel}}$$

because then, $n < 0$ and $\mu_{rel} < 0$ will imply that the characteristic impedance of the medium $\eta = \eta_0 \mu_{rel}/n$ will be positive, which, as we will see later, implies that the energy flux of a wave is in the same direction as the direction of propagation.

In media with simultaneously negative permittivity and permeability, $\varepsilon < 0$ and $\mu < 0$, the refractive index must be negative [15]. To see this, we consider a uniform plane wave propagating in a lossless medium:

$$
\begin{aligned}
E_x(z, t) &= E_0 e^{(j\omega t - jkz)} \\
H_y(z, t) &= H_0 e^{(j\omega t - jkz)}.
\end{aligned}
\tag{2.27}
$$

Then, Maxwell's equations require the following relationships, which are equivalent to Faraday's and Ampere's laws, respectively:

$$kE_0 = \omega\mu H_0$$

$$kH_0 = \omega\varepsilon E_0$$

or

$$
\eta = \frac{E_0}{H_0} = \frac{\omega\mu}{k} = \frac{k}{\omega\varepsilon}
\tag{2.28}
$$

$$\Rightarrow k^2 = \omega^2 \varepsilon\mu.$$

Because the medium is lossless, k and η will be real and the time-averaged Poynting vector, which points in the z-direction, will be

$$
\mathbf{p}_z = \frac{1}{2}\mathrm{Re}\left[E_0 H_0^*\right] = \frac{1}{2\eta}|E_0|^2 = \frac{1}{2}\eta|H_0|^2.
\tag{2.29}
$$

If we require that the energy flux is toward the positive z-direction, that is, $\mathbf{p}_z > 0$, then we must have $\eta > 0$. Because μ and ε are negative, Equation (2.28) implies that k must be negative, $k < 0$, in order for the ratio $\eta = \omega\mu/k$ to be positive. Thus, in solving $k^2 = \omega^2\mu\varepsilon$, we must choose the negative square root:

$$
k = -\omega\sqrt{\varepsilon\mu}.
\tag{2.30}
$$

The refractive index n may be defined through $k = k_0 n$, where $k_0 = \omega(\mu_0\varepsilon_0)^{0.5}$ is the free-space wave number. Thus, we can express n in terms of the relative permittivity and permeability:

$$
n = \frac{k}{k_0} = -\sqrt{\frac{\varepsilon}{\varepsilon_0}\frac{\mu}{\mu_0}} - \sqrt{\varepsilon_{rel}\mu_{rel}}.
$$

Writing $\varepsilon = -|\varepsilon|$ and $\mu = -|\mu|$, we have for the medium impedance

$$
\eta = \frac{\omega\mu}{k} = \frac{-\omega|\mu|}{-\omega\sqrt{|\mu\varepsilon|}} = \sqrt{\frac{\mu}{\varepsilon}},
\tag{2.31}
$$

Figure 2.15. (a) A common refractive medium. (b) A metamaterial.

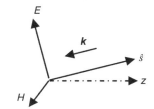

Figure 2.16. Propagation direction.

which can be written also as follows, where $\eta_0 = (\mu_0/\varepsilon_0)^{0.5}$:

$$\eta = \eta_0 \frac{\mu}{\mu_0 n} = \eta_0 \frac{\varepsilon_0 n}{\varepsilon}. \tag{2.32}$$

Thus, in metamaterials, the wave vector k and the *phase velocity* $v_{ph} = \omega/k = c_0/n$ will be negative, pointing in the opposite direction than the Poynting vector. For lossless metamaterials, the *energy transport velocity* v_{en}, which is in the direction of the Poynting vector, coincides with the group velocity v_g. Thus, $v_g = v_{en} > 0$, while $v_{ph} < 0$. A diagram of the positive and negative refractive index is shown in Figure 2.15.

Figure 2.15a is a light beam passing from air or a vacuum into a common refractive medium such as glass, water, or quartz is bent at the surface boundary so its path inside the material is more nearly perpendicular to the surface than its path outside the material. Figure 2.15b shows the extent of the bending depends on the angle at which the ray strikes the boundary and also on the index of refraction of the medium. All common transparent materials have positive indices of refraction. A metamaterial bends an incident ray so its internal direction is reversed.

These and other consequences of $n < 0$, such as the reversal of the Doppler and Cherenkov effects and the reversal of the field momentum, have been discussed in References 15 and 18.

- Arbitrary Propagation

If the propagation is along an arbitrary direction defined by a unit vector $\hat{s}$ (i.e., a rotated version of $\hat{z}$; Fig. 2.16), then we may define the wave vector by $\boldsymbol{k} = k\hat{s}$, with k to be determined, and look for solutions of Maxwell's equations of the form

$$E(r, t) = E_0 e^{(j\omega t - jk \cdot r)}$$
$$H(r, t) = H_0 e^{(j\omega t - jk \cdot r)}.$$

(2.33)

Gauss's laws require that the constant vectors E_0, H_0 be transverse to k or $\hat{s}$; that is, $\hat{s} \cdot E_0 = \hat{s} \cdot H_0 = 0$. Then, Faraday's and Ampere's laws require that

$$H_0 = \frac{1}{\eta}(\hat{s} \times E_0),$$

$$\eta = \frac{\omega\mu}{k} = \frac{k}{\omega\varepsilon} \Rightarrow k^2 = \omega^2 \mu\varepsilon$$

(2.34)

with a Poynting vector:

$$\mathbf{p} = \frac{1}{2}\text{Re}\left[E_0 \times H_0^*\right] = \hat{s}\frac{1}{2\eta}|E_0|^2.$$

(2.35)

Thus, if $\mathbf{p}$ is assumed to be in the direction of $\hat{s}$, then we must have $\eta > 0$, and therefore, k must be negative as in Equation (2.30). It follows that the wave vector $k = k\hat{s}$ will be in the opposite direction of $\hat{s}$ and $\mathbf{p}$.

Equation (2.34) implies that the triplet $\{E_0, H_0, \hat{s}\}$ is still a right-handed vector system, but $\{E_0, H_0, k\}$ will be a left-handed system. Therefore, such media is also named left-handed media [15].

In a lossy negative-index medium, the permittivity and permeability are complex-valued, $\varepsilon = \varepsilon_r - j\varepsilon_i$ and $\mu = \mu_r - j\mu_i$, with negative real parts ε_r, $\mu_r < 0$ and positive imaginary parts ε_i, $\mu_i > 0$. Equation (2.28) remains the same and will imply that k and η will be complex-valued. Letting $k = \beta - j\alpha$, the fields will be attenuating as they propagate:

$$E_x(z, t) = E_0 e^{-\alpha z} e^{j\omega t - j\beta z}$$

$$H_y(z, t) = H_0 e^{-\alpha z} e^{j\omega t - j\beta z},$$

and the Poynting vector $\mathbf{p}$ will be given by

$$\mathbf{p}_z = \frac{1}{2}\text{Re}\left[E_x(z)H_y^*(z)\right]$$

$$= \frac{1}{2}\text{Re}\left(\frac{1}{\eta}\right)|E_0|^2 e^{-2\alpha z}$$

$$= \frac{1}{2}\text{Re}(\eta)|H_0|^2 e^{-2\alpha z}.$$

(2.36)

The refractive index is complex-valued, $n = n_r - jn_i$, and is related to k through $k = k_0 n = k_0(n_r - jn_i)$. If k is represented as $\beta - j\alpha$, we have $\beta = k_0 n_r$ and $\alpha = k_0 n_i$. Thus, the conditions of negative phase velocity ($\beta < 0$), field attenuation ($\alpha > 0$), and positive power flow can be written as follows:

$$n_r < 0,$$
$$n_i > 0, \tag{2.37}$$
$$\mathrm{Re}(\eta) > 0.$$

REFERENCES

[1] S. J. Orfanidis, Electromagnetic Waves and Antennas, Rutgers University Press, Piscataway, NJ, 2008.

[2] J. C. Maxwell, A Treatise on Electricity and Magnetism, Oxford University Press, Oxford, 1983.

[3] J. R. James, "Theoretical Investigation of Cylindrical Dielectric Rod Antennas," Proceedings of the Institution of Electrical Engineers Vol. 114, 1967, pp. 309–319.

[4] R. S. Elliott, Antenna Theory and Design, Series on ElectromagneticWave Theory, IEEE Press, New York, 2006.

[5] K. Huang and Z. Wang, "V-Band Patch-Fed Rod Antennas for High Data-Rate Wireless Communications," IEEE Transactions on Antennas and Propagation Vol. 54, January 2006, pp. 297–300.

[6] F. Colomb, K. Hur, W. Stacey, and M. Grigas, "Annular Slot Antennas on Extended Hemispherical Dielectric Lenses," Antennas and Propagation Society International Symposium, 1996, AP-S Digest, 3, July 1996, 2192–2195, 1996.

[7] S. Kobayashi, R. Mittra, and R. Lampe, "Dielectric Tapered Rod Antennas for Millimeter-Wave Applications," IEEE Transactions on Antennas and Propagation Vol. 30, 1982, pp. 54–58.

[8] X. Wu, G. V. Eleftheriades, and T. E. van Deventer-Perkins, "Design and Characterization of Single- and Multiple-Beam mm-Wave Circularly Polarized Substrate Lens Antennas for Wireless Communications," IEEE Transactions on Microwave Theory and Techniques Vol. 49, No. 3, March 2001, pp. 431–441, 2004.

[9] X. Wu and G. V. Eleftheriades, "Two-Lens and Lens-Fed Reflector Antenna Systems for mm-Wave Wireless Communications," IEEE Antennas and Propagation Society International Symposium Vol. 2, 2000, pp. 660–663.

[10] J. R. Costa, E. B. Lima, and C. A. Fernandes, "Antenna Phase Center Determination from Amplitude Measurements Using a Focusing Lens," IEEE Antennas and Propagation Society International Symposium (APSURSI), July 2010, pp. 1–4, 2010.

[11] T. Manabe et al., "Polarization Dependence of Multipath Propagation and High-Speed Transmission Characteristics of Indoor Millimeter-Wave Channel at 60 GHz," IEEE Transactions on Vehicular Technology Vol. 44, No. 2, May 1995, pp. 268–274.

[12] A. C. Ludwig, "The Definition of Cross Polarization," IEEE Transactions on Antennas and Propagation Vol. AP-21, No. 1, January 1973, pp. 116–119.

[13] L. D. Landau, E. M. Lifshitz, and L. Pitaevskii, Electrodynamics of Continuous Media, 2nd ed., Elsevier Science, Burlington, MA, 1985.

[14] J. B. Pendry, "Magnetism from Conductors and Enhanced Nonlinear Phenomena," IEEE Transactions on Microwave Theory and Techniques Vol. 47, 1999, pp. 2075–2084.

[15] V. G. Veselago, "The Electrodynamics of Substances with Simultaneously Negative Values of ε and μ," Soviet Physics Uspekhi Vol. 10, 1968, p. 509.

[16] D. R. Smith, et al., "Composite Medium with Simultaneously Negative Permeability and Permittivity," Physical Review Letters Vol. 84, 2000, p. 4184.

[17] W. J. Padilla, D. R. Smith, and D. N. Basov, "Spectroscopy of Metamaterials from Infrared to Optical Frequencies," Journal of the Optical Society of America Vol. B-23, 2006, p. 404.

[18] R. A. Depine and A. Lakhtakia, "A New Condition to Identify Isotropic Dielectric-Magnetic Materials Displaying Negative Phase Velocity," Microwave and Optical Technology Letters Vol. 41, 2004, p. 315.

3

POLYROD ANTENNAS

Kao-Cheng Huang

Polyrod antennas have been used extensively as directive antennas for many years. They are "end-fire" radiators and act in a similar manner to electromagnetic waveguides with considerable power radiating through the wall of the polyrods. A considerable number of theoretical and experimental studies for this type of antenna have been published, these concerning microwave and millimeter wave frequencies [1]. This type of antenna has the attractive features of light weight, low cost, no inherent conductor loss, high radiation efficiency, high directivity, and wide bandwidth. Also, in common with more conventional lens antennas, the polyrod structure is not too sensitive to dimensional tolerances. Polyrods are encountered in a variety of simple geometries, among which are hemispherical, circular cylinders and those with a rectangular cross section. These are readily available and can be easily fabricated. They are very compact in size when operating at a high frequency and with a large dielectric constant ε_r, which could be as high as 30–100. The dimensions scale with free-space wavelength and dielectric constant in the usual relationship and so are in the order of $\lambda_0/\sqrt{\varepsilon_r}$.

Impedance bandwidth varies over a wide range according to the resonator parameters. It can be as small as a few percent with high constant materials, or over 20%

Modern Lens Antennas for Communications Engineering, First Edition. John Thornton and Kao-Cheng Huang.
© 2013 Institute of Electrical and Electronics Engineers. Published 2013 by John Wiley & Sons, Inc.

with smaller ε_r used in conjunction with certain geometries and resonant modes. Different far-field radiation patterns are supported. For a given polyrod geometry, the radiation patterns can be made to change by exciting different resonant modes. The feeding mechanism is simple, flexible, and easily controlled. Different kinds of transmission line feeds are suitable for coupling to the polyrod radiator: conducting probes, microstrip lines, and microstrip slots are frequently used.

In light of these advantages, we present the design considerations, properties, and techniques of polyrods in this chapter, which is divided into the following sections. Section 3.1 further introduces the polyrod as a resonator. Section 3.2 describes its properties as a radiator and then tapered polyrods. Section 3.3 presents the basic features of a microstrip patch-fed dielectric antenna, while Section 3.4 analyzes arrays of polyrod elements. Finally, in Section 3.5, a multibeam polyrod array is introduced, which has wide coverage.

3.1 POLYRODS AS RESONATORS

Polyrod resonators appeared in the 1970s in work that led to the miniaturization of active and passive microwave components, such as oscillators and filters [2]. In a shielded environment, the resonators built with polyrod resonators can reach the unloaded quality (Q) factor of 2×10^4 at frequencies of 20 GHz.

The principle of operation of the polyrod resonator can best be understood by studying the propagation of electromagnetic waves on a polyrod waveguide [2]. The mathematical description [3] and the experimental verification [4] of the existence of these waves have been known for a considerable time. Their wide application was prompted by the introduction of optical fibers in communications systems.

One of the attractive features of a polyrod resonator antenna is that it can assume any one of a number of simple shapes, the most common being ones with circular or rectangular cross sections, as shown in Figure 3.1. Over the years, the frequency range of interest for many systems has gradually progressed upward to the millimeter and near-millimeter range (100–300 GHz). At these frequencies, the conductor loss of metallic antennas becomes severe and the efficiency of such antennas can be reduced significantly. Conversely, the only loss for a polyrod resonator antennas is that due to the imperfect dielectric material (the loss tangent of the material), which can be very small in practice. After the cylindrical polyrod resonator antenna had been studied [5], the rectangular and hemispherical variants were subsequently investigated [6]. The work created the foundation for future investigations of yet further geometric variants. Other shapes were studied, including cylindrical-ring [7] and spherical-cap polyrod resonator antennas. It was found that polyrod resonator antennas operating at their fundamental modes radiate like a magnetic dipole, independent of their shapes.

When a polyrod resonator is not entirely enclosed by a conductive boundary, it can radiate, and so it becomes an antenna. An early polyrod resonator antenna was successfully built and described in Reference 5, while the rigorous numerical solution was published in Reference 8. Review treatments of polyrod resonator antennas can be found in References 9 and 10.

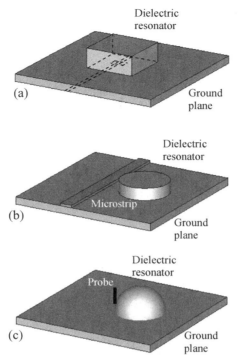

Figure 3.1. Typical polyrod resonator antennas and feeding mechanisms: (a) aperture feed, (b) microstrip feed, and (c) probe feed.

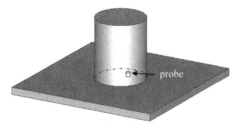

Figure 3.2. Polyrod resonator antenna fed with a coaxial probe.

Shown in Figure 3.2 is a polyrod resonator element, which is placed on a ground plane, and a short electric probe penetrates the plane into the resonator. The probe is located off the center of the resonator, close to the perimeter of the cylindrical resonator. The radiation occurs mainly in the broadside direction (i.e., radially) and it is linearly polarized.

As with all bounded systems, the field distribution within the structure is defined by the mode. The mode of propagation is a solution to the wave equation, which

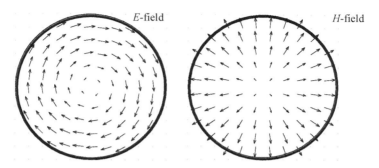

Figure 3.3. Mode TE_{01} on a polyrod. Left: E-field, right: H-field.

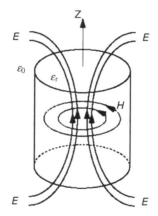

Figure 3.4. Mode TM_{01} on a polyrod showing the E-field and the H-field.

satisfies both the boundary conditions and the excitation (feeding) method. Some of the lowest modes of propagation on polyrod waveguides are shown in Figures 3.3–3.5.

For the *TE*, *TM*, and *HEM* modes, the first index denotes the number of full-period field variations in the azimuthal direction, and the second index the number of radial variations. When the first index is equal to zero, the electromagnetic field is circularly symmetric. In the cross-sectional view, the field lines can be either concentric circles (e.g., the E-field of the TE_{01} mode) or a radial straight line (e.g., the H-field of the same mode). For higher modes, the pure transverse electric or transverse magnetic fields cannot exist, so that both electric and magnetic field must have nonvanishing longitudinal components. Such modes are called hybrid electromagnetic (*HEM*), the lowest of them being HEM_{11}.

The fields of a cylindrical polyrod can be expressed in terms of Bessel functions, which determine the wavelength and the propagation velocity of these waves. When

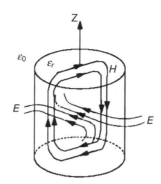

Figure 3.5. Mode HEM_{11} on a polyrod.

only a truncated section of the polyrod waveguide is used, one obtains a resonant cavity in which standing waves appear and which gives rise to the term polyrod resonator. The resonant mode $TE_{01\delta}$ is most often used in shielded microwave circuits. In classical waveguide cavities, the third index is used to denote the number of half-wavelength variations in the axial direction of the waveguide. In the polyrod, the third index, δ, denotes the fact that the resonator is shorter than one-half wavelength. The actual height depends on the relative dielectric constant of the resonator and the substrate, and on the proximity to the top and bottom conductor planes. Since the numerical value of δ is rarely used, this index is usually ignored, so that the polyrod resonator is often specified by two indices only.

The numerical analysis of the polyrod resonator antenna started as an attempt to determine the natural frequencies of the isolated structure, free of other scattering objects in its vicinity, and also without any excitation mechanism. It was found that the resonant frequencies are complex-valued, given by

$$f_{m,n} = \sigma_{m,n} + j\omega_{m,n}, \tag{3.1}$$

where

σ represents the in-phase (real) component (which is the lossy component) and ω is the out-of-phase component (imaginary) with respect to the excitation.

Each particular solution corresponds to a resonant m, n-type mode that satisfies all the boundary and continuity conditions. For rotationally symmetric resonators, subscript m denotes the number of azimuthal variations, and subscript n denotes the order of appearance of modes in the growing frequency direction.

The fact that the resonant frequency has a nonvanishing real part signifies that such a mode would oscillate in an exponentially decaying manner, if it was initially excited by an abrupt external stimulus. The ratio of the real to the imaginary parts of the resonant frequency is the radiation Q factor of the mode:

$$Q_r = -\frac{\omega_{m,n}}{2\sigma_{m,n}}. \tag{3.2}$$

The negative sign comes from the fact that all passive circuits have their natural frequencies located on the left-half complex plane, so $\sigma_{m,n}$ is itself a negative number. The natural frequencies and the radiation Q factors of the modes are given as follows [2]:

- $TE_{01\delta}$ mode

$$f_0 = \frac{2.921 c\varepsilon_r^{-0.465}}{2\pi A}\left[0.691 + 0.319\left(\frac{A}{2H}\right) - 0.035\left(\frac{A}{2H}\right)^2\right] \tag{3.3}$$

$$Q = 0.012\varepsilon_r^{1.2076}\left[5.270\left(\frac{A}{2H}\right) + 1106.188\left(\frac{A}{2H}\right)^{0.625} e^{-1.0272\left(\frac{A}{2H}\right)}\right] \tag{3.4}$$

- $HE_{11\delta}$ mode

$$f_0 = \frac{2.735 c\varepsilon_r^{-0.436}}{2\pi A}\left[0.543 + 0.589\left(\frac{A}{2H}\right) - 0.050\left(\frac{A}{2H}\right)^2\right] \tag{3.5}$$

$$Q = 0.013\varepsilon_r^{1.202}\left[2.135\left(\frac{A}{2H}\right) + 228\left(\frac{A}{2H}\right)e^{-2\left(\frac{A}{2H}\right)+0.11\left(\frac{A}{2H}\right)^2}\right] \tag{3.6}$$

- $TM_{01\delta}$ mode

$$f_0 = \frac{2.933 c\varepsilon_r^{-0.468}}{2\pi A}\left\{1 - \left[0.075 - 0.05\left(\frac{A}{2H}\right)\right]\left[\frac{\varepsilon_r - 10}{28}\right]\right\}$$
$$\left\{1.048 + 0.377\left(\frac{A}{2H}\right) - 0.071\left(\frac{A}{2H}\right)^2\right\} \tag{3.7}$$

$$Q = 0.009\varepsilon_r^{0.888} e^{0.04\varepsilon_r}\left\{1 - \left[0.3 - 0.2\left(\frac{A}{2H}\right)\right]\left[\frac{38 - \varepsilon_r}{28}\right]\right\}$$
$$\left\{9.498\left(\frac{A}{2H}\right) + 2058.33\left(\frac{A}{2H}\right)^{4.322} e^{-3.5\left(\frac{A}{2H}\right)}\right\} \tag{3.8}$$

For given dimensions and dielectric constant, the numerical solutions determine both the resonant frequency and the radiation Q factor. Such computed data can be fitted to convenient analytic expressions [11]. For instance, the resonant frequency of the HEM_{11} mode of an isolated polyrod resonator radiator of radius a and height h can be approximated by the following expression:

$$k_0 a = (1.6 + 0.513x + 1.329x^2 - 0.575x^3 + 0.088x^4)/\varepsilon_r^{0.42}, \tag{3.9}$$

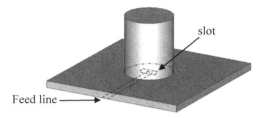

Figure 3.6. Polyrod resonator antenna with a microstrip-slot excitation.

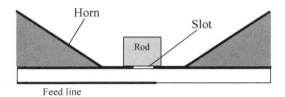

Figure 3.7. Aperture-coupled rectangular polyrod antenna with a quasi-planar horn.

where k_0 is the free-space propagation constant and $x = a/h$.

Similarly, the values of Q_r for the same mode can be calculated from

$$Q_r = x\varepsilon_r^{1.2}(1.01893 + 2.925e^{-2.08x(1-0.08x)}).$$ (3.10)

An alternative way of exciting the HEM_{11} mode in the polyrod resonator antenna is by the microstrip-slot mechanism shown in Figure 3.6. Instead of a coaxial line, the feeding is done by a microstrip line that runs below the ground plane. There is a narrow slot (aperture) in the ground plane (which is the upper layer here) for coupling the microstrip to the polyrod above it.

There is a gain enhancement technique available for the polyrod by using a surface mounted quasi-planar horn. The geometry of this antenna is shown in Figure 3.7. In the polyrod–horn structure, the aperture-coupled polyrod now works as a feed to the surface mounted horn antenna. Note that the rectangular polyrod is located over the center of the rectangular slot (aperture feed) in the ground and is excited by a 50-Ω microstrip line feed. If the surface mounted horn is made of thin copper or aluminum sheet and is supported on a foam structure, the gain can be further improved.

3.2 THE POLYROD AS A RADIATOR

Polyrod antennas are of considerable interest as medium gain antennas in their own right and also as radiating elements in arrays. Applications include terahertz imaging and high-throughput communications where an adequate signal-to-noise ratio is sought and which might be otherwise difficult to achieve owing to deficiencies in sources and detectors at these high frequencies. The approach to antenna design has commonly been to treat the polyrod as a linear array of equivalent sources, these arising from radiation from discontinuities experienced by the surface wave traveling along the antenna.

Polyrod surface-wave antennas are good directional radiators in the end-fire direction and those with circular cross section have been among the first to be investigated analytically and experimentally [12, 13]. However, the application of these antennas has been limited due to their relatively low gain. Recent commercial interest in developing millimeter wave dielectric circuits created the additional need for low-cost antennas that can be easily integrated into an entire system [13].

Thus, dielectric antennas of rectangular cross section that are compatible with the dielectric waveguides of a millimeter wave integrated circuit show practical advantages and are discussed next. Presented in this section are experimental results for tapered polyrod antennas designed for low sidelobes and for maximum gain.

Polyrod antennas have been used as end-fire radiators for many years [14–20]. Experimental studies have been conducted both at microwave and millimeter wave frequencies [17–20]. Despite the extensive use of these antennas, no exact design procedure exists for them [20]. Theoretical methods usually involve simplifications and only provide general design guidelines [14–16].

The radiation behavior of the polyrod antenna can be explained by the so-called discontinuity-radiation concept [14], in which the antenna is regarded as an array composed of two effective sources at the feed and free ends of the polyrod. Part of the power excited at the feed is converted into guided-wave power and is transformed into radiated power in free space at the "open" end.

The remaining power is converted into unguided-wave power radiating near the feed end. Thus, the directivity of the polyrod antenna is characterized by the directivities generated by these two effective sources. However, it is complicated by the problem of quantitatively computing the radiation fields generated from the discontinuities at the feed and free ends.

On the other hand, the polyrod has received much attention in the waveguide analysis as well as the antenna analysis since the knowledge of the eigenmode is of fundamental importance in the design of a dielectric waveguide circuit. Numerous approaches have been proposed for this important issue [21–26]. The methods developed in References 24–26 are based on Yee's mesh, the use of which has the advantage that the obtained eigenmode fields can directly be used in the finite-difference time-domain (FDTD) method [27, 28].

The field near the polyrod can be decomposed into the guided and unguided waves. Using the obtained solutions, we can calculate a feed pattern corresponding to the directivity generated by the effective source at the feed end and a terminal pattern corresponding to that at the free end. Superposition of the feed and terminal patterns gives rise to the aggregate radiation pattern of the polyrod antenna. The gain of a long polyrod antenna is similarly calculated by superposing the feed and terminal patterns. The details of the numerical analysis can be found in Reference 28.

Figure 3.8 shows the configuration of a polyrod antenna fed by a rectangular waveguide with a planar ground plane. The polyrod is made of Teflon (a registered trademark of polytetrafluoroethylene [PTFE]). It is assumed that the metallic waveguide

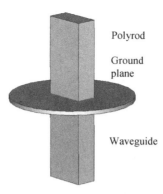

Figure 3.8. Overall geometry of a rectangular cross-section polyrod antenna and waveguide.

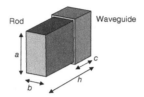

Figure 3.9. Rectangular polyrod (a = 4.4 mm, b = 2.2 mm, c = 3 mm, h = 8 mm).

and the ground plane are perfectly conducting, and that the polyrod is a lossless medium [29–31]. The basic dimensions of such a polyrod antenna are shown in Figure 3.9. Here, the dimensions of the base area determine the resonant frequency. In this example, the width b is half that of the length a; that is, $a = 2b$. Then, if this ratio is altered by changing the dimension b, the resonant frequency changes according to the curves shown in Figure 3.10.

The gain of the polyrod antenna increases as the height increases. Figure 3.11 shows the gain patterns for polyrod height h = 6, 8, and 12 mm, respectively. With the increase in gain, the form of the sidelobes can also vary quite strongly, and so careful tuning should be expected to be necessary to arrive at a favorable compromise.

Eventually, of course, the gain reaches a maximum value and does not change too much with further increases in the height h: This effect is shown in Figure 3.12. Accordingly, it is worth knowing the maximum gain that can be expected for the polyrod and at what maximum dimension. So from this effect, we can assess the performance expected from a compact polyrod antenna.

3.2.1 Tapered Polyrod Antenna

The tapered polyrod antenna can take any of the configurations shown in Figure 3.13. The cross section may be rectangular or circular, although Figure 3.13 shows only the former case. Here, the taper may occur in one cross-sectional dimension or in both and

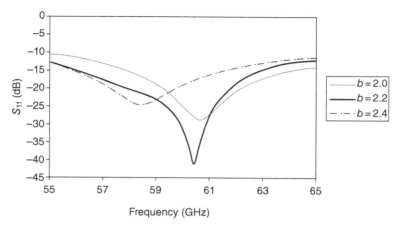

Figure 3.10. Resonant frequency of a polyrod antenna as the rectangular cross section varies.

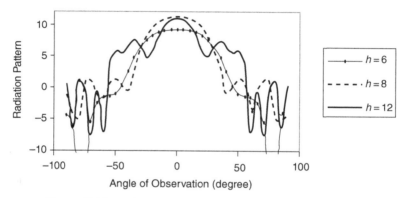

Figure 3.11. Radiation patterns for different polyrod heights.

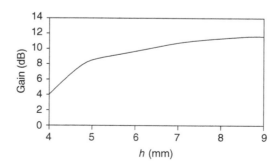

Figure 3.12. Gain curve versus height of the polyrod antenna in Figure 3.9 (*a*, *b*, and *c* remain fixed).

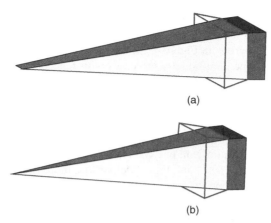

(a)

(b)

Figure 3.13. Examples of polyrod antennas with tapered cross sections: (a) single-axis taper to a line and (b) two-axis taper to a point.

may be either symmetric or asymmetric. Millimeter wave polyrod antennas will usually have a rectangular cross section and a linear taper, as indicated in Figure 3.13a, this being preferable for reasons of simplicity. Such antennas have been designed for the submillimeter wave region [32].

In most experimental investigations reported in the literature, the antennas are fed by metal waveguides [33]. In this case, radiation from the feed aperture can have a strong effect on the pattern shape and sidelobe level, and an appropriate feed arrangement must be used to minimize transition effects [34]. If the antenna is fed by dielectric waveguides made from the same material as the polyrod, the transition region will cause little radiation and a smooth pattern can be expected. An interesting case is where the apex angle is small, giving rise to a quite slender structure.

It should be noted that in the absence of the short horn in the proximity of the feed—polyrod transition, this feed point would radiate quite heavily and so contribute to sidelobes in the far-field pattern. (The envelope of these sidelobes arises from those of the aperture at the abrupt termination of the waveguide.)

A polyrod tapered in only the E-plane as shown in Figure 3.13a is both easier to manufacture and mechanically stronger than its counterpart in Figure 3.13b, which tapers in both the E- and H-planes, yet shows no sacrifice in antenna characteristics and can even sometimes show a little improvement. Conversely, a polyrod tapered in only the H-plane will always be worse than one tapered in both planes.

The second type of polyrod antenna, shown in Figure 3.13b, is also linearly tapered to a point in both the E- and H-planes. Some experimental data for this type are offered by Kobayashi et al. [1], where measurements showed a sidelobe level lower than −25.5 dB from the main beam and a gain of 17.0 dBi, which is slightly lower than that of the feed horn by itself at 17.3 dBi. Its half-power beamwidth is 30.0°, which is larger than that of the feed horn (26.9°); however, the 26-dB beamwidth of this antenna is only 72° compared with 115° for the horn. Small sidelobe characteristics and steep sidelobe roll-off are features of a relatively short tapered polyrod antenna with a feed horn.

For the tapered polyrods in Figure 3.13, the directivity is determined primarily by the antenna length L. The gain initially increases with L according to the relation [13]

$$G = \rho\left(\frac{L}{\lambda_0}\right), \tag{3.11}$$

where ρ is approximately 7 for long antennas having a linear taper.

A problem with maximum-gain polyrod antennas is that they have a comparatively small pattern bandwidth ($\pm10\%$) and high sidelobes no more than -6 to -10 dB down from the main peak. Appropriate design of the body taper will solve these problems, though at the expense of a reduced gain [13]. Most tapered polyrod antennas can be expected to have gain and beamwidth values in a compromise region. Experimental evidence in Reference 1 shows, in particular, that antennas with a linear taper can be designed to have a gain not significantly lower than that of maximum-gain antennas, while their pattern quality is substantially better with sidelobes at less than -20 dB. Since these antennas have the additional advantage of structural simplicity, they should be of more interest for millimeter wave applications than maximum-gain antennas.

In the experiments of interest here, care should be taken to minimize parasitic radiation from the feed-to-antenna transitions [1]. Further design recommendations are the following:

(a) Antennas of moderate length ($3 \lambda_0$ to $10 \lambda_0$) tend to produce a comparatively broad main beam whose sides, however, drop down rapidly toward the first zero.

(b) The beamwidth and sidelobe level decrease monotonically with the increasing axial length of the antenna. However, if an optimum length is exceeded, main-beam distortions occur, leading to a significantly increased sidelobe level and, eventually, a pattern breakup.

(c) The beamwidth of the E- and H-plane patterns is essentially the same. By careful design, the sidelobe level can be kept below -25 dB. Figure 3.14 shows the pattern of a pyramidal antenna of length $L = 5.5 \lambda_0$ and permittivity $\varepsilon_r = 2.1$ (Teflon). The antenna was fed by a metal waveguide with a horn launcher.

Nonlinear or curvilinear tapered polyrods are other possible variants as can be seen in Figure 3.15. Here, the dielectric polyrod is excited by a ring slot, which is implemented in a ground plane. Both polyrods have diameters that gradually narrow before flaring at the end of the rod. The effect of this oscillatory profile is to detach the surface wave from the polyrod and encourage radiation, so increasing the antenna gain. The calculated radiation pattern for the dielectric polyrod in Figure 3.15a at the center of the frequency band is shown in Figure 3.16.

A comparison between a tapered polyrod antenna and an aperture antenna (e.g., a parabolic dish) shows that these antennas provide approximately the same directivity gain and sidelobe performance if their dimensions are related by

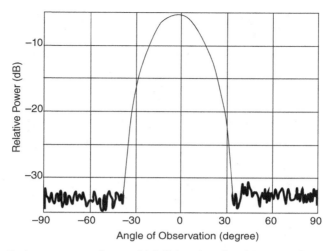

Figure 3.14. *H*-plane pattern of pyramidal dielectric-polyrod antenna fed by metal wave-guide with horn launcher (©1982 IEEE [1]).

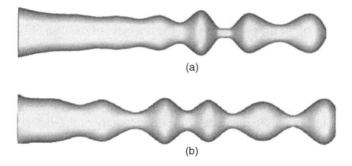

Figure 3.15. Two nonlinear profiled dielectric polyrod antennas (© 2009 IEEE [35]).

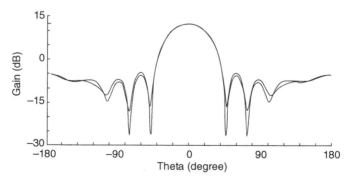

Figure 3.16. *E*-plane radiation pattern for dielectric polyrod in Figure 3.15a (© 2009 IEEE [35]).

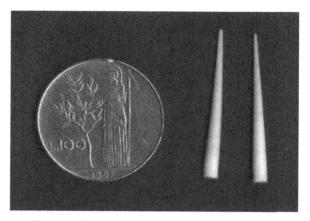

Figure 3.17. A pair of 61-GHz polyrod antennas.

$$\frac{L}{\lambda_0} = \left(\frac{D}{\lambda_0}\right)^2, \tag{3.12}$$

where L is the length of the polyrod and D the diameter of the dish [13].

3.3 PATCH-FED CIRCULAR POLYROD

As can be seen in Figure 3.17, a polyrod can act as a guide for electromagnetic waves. However, depending on the magnitude of the discontinuity of the dielectric constant at the boundary, a considerable amount of radio frequency power can propagate through the surface of the polyrod and is radiated into free space. This radiation mechanism is exploited in the design of the polyrod antenna. It can be represented as a patch and a waveguide, that is, consisting of a cylindrical part and a tapered part. The polyrod is fed by the patch, which is energized by a microstrip line, this in turn typically connected to a coaxial line or adaptor. These antenna configurations can be easily built and integrated with other millimeter wave functional modules or planar circuits.

A patch antenna itself produces the TM_{010} fundamental mode. When a polyrod is put onto the patch and is surrounded by a metallic waveguide, complex mode excitations are generated. While the antenna is radiating, energy from the patch is transferred to the tapered polyrod through a small cylindrical rod and a circular waveguide. To take an example, the height of the short cylindrical rod is set to 3 mm, while that of the circular waveguide is 7 mm (Fig. 3.17). The cylindrical polyrod and the circular waveguide act as a mode converter, which mainly excites the TE mode. Higher modes are suppressed by selecting the height of the waveguide and the diameter of the polyrod.

If the diameter of the cylindrical polyrod antenna is smaller than a quarter wavelength, only a small amount of the energy is kept inside the dielectric rod, which shows little guiding effect on the wave. Also, in this case, the phase velocity in the polyrod is nearly the same as in free space. In contrast, when the diameter increases to the order of one wavelength, most of the electromagnetic energy is contained by the polyrod and the phase velocity becomes approximately the same as it would be in a boundless dielectric material.

The dominant mode on the tapered polyrod is HE_{11}, as encountered in a circular cross-section waveguide. The lowest mode in such a circular waveguide is TE_{11}, which propagates when the diameter of the guide exceeds $0.58\lambda/\sqrt{\varepsilon_r}$, where λ is the free-space wavelength and ε_r is the relative permittivity [36]. Thus, for a polyrod terminated by a circular waveguide, the guide diameter must be at least 0.37λ to allow the HE_{11} mode to propagate in the metal tube.

The diameter of a PTFE polyrod antenna (where $\varepsilon_r = 2.1$) would be 3 mm for a frequency of 61 GHz and its height 3 mm at the base, as shown in Figure 3.17. The upper part of the polyrod is tapered linearly to a terminal with 0.7-mm diameter and 30-mm height in order to achieve a high antenna gain. The tapered polyrod can be treated as an impedance transformer and it reduces the reflection caused by an abrupt discontinuity [1, 13].

While being fed by a patch, the diameter of the polyrod antenna also matches to the small waveguide at its base. The inner diameter of the waveguide is 3 mm, which is the same as the diameter of the cylindrical polyrod. This waveguide conducts electromagnetic energy between the polyrod antenna and the patch antenna. This constraint decreases the radiation leakage in unwanted directions and reduces the sidelobes in the far-field radiation pattern. In addition, this waveguide can assist with good mechanical alignment between the polyrod antenna and the patch antenna, which lends this design toward use in mass production. It does not have to touch ground plane when a thin substrate is used. The waveguide can be fixed onto the substrate by means of epoxy or a mechanical fixture.

The heights of the cylindrical and tapered polyrods in the above examples follow Zucker's design rules [13]. As we have seen as a general rule, when the height of the polyrod antenna is reduced, the gain will reduce and the half-power beamwidth sees an increase.

The patch antenna used here is fabricated on Rogers *RT/Duroid* 5880 substrate with ε_r 2.2 and thickness 110 microns. The resulting patch size patch is 0.7×0.7 mm approximately, with the target center frequency at 61 GHz. The substrate material has a similar permittivity to the PTFE polyrod and, as a fortunate consequence, the guided wave at the boundary between them suffers minimal mismatch loss that would otherwise arise from a more abrupt impedance change.

The polyrod antenna with circular cross section has a symmetrical shape and therefore it generates the same energy in both right-hand circular polarization (RHCP) and left-hand circular polarization (LHCP) and so gives rise to linear polarization. If an asymmetry is deliberately introduced into the designed, the rod will radiate more energy in one circular polarization than the other. For instance, the top surface of the rod can be machined with an oblique slope at the upper face instead of the flat surface. Assuming that the feed is a truncated square patch with a microstrip line extending from the $-y$-axis toward the $+y$-axis direction, these different rod upper surfaces correspond to the different polarizations shown in Figure 3.18.

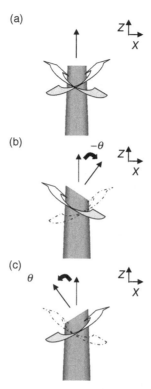

Figure 3.18. Polarization changes for different profiles at the top surface of a polyrod antenna. (a) Symmetry, (b) LHCP, and (c) RHCP (© 2006 IEEE [37]).

In Figure 3.18, the solid arrows show the polarization sense of the dominant electrical field, while the dotted arrows show that of the suppressed component, this being of the opposite sense.

Recalling that the polyrod antenna is a directional antenna and radiates along the central axis of the rod, the direction of the main beam can be easily adjusted by tilting this central axis. Figure 3.19 shows an example of using the polyrod antenna in different radiation directions. The shape at the interface to the feeding waveguide should here also be modified to accommodate the tilt of the polyrod.

Figure 3.20 illustrates the dependency of radiation pattern on the polyrod height, where values were taken at 10, 18, and 30 mm, respectively. The radiation patterns shown are in each of the three cases simulated at 61 GHz. As we have seen, increasing rod height leads to gain increase and corresponding reduction in the half-power beamwidth. A maximum of 17-dBi gain occurs for a height of 22 mm (the antenna gain becomes saturated) and further increases in height see the gain dip down. Distortion of the main beam occurs, which is manifested in the presence of sidelobes, which are also seen to vary quite significantly with changes in H.

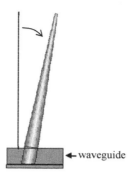

Figure 3.19. Geometry of a beam-tilting polyrod antenna: The beam direction changes as the rod is tilted away from the broadside direction (© 2006 IEEE [37]).

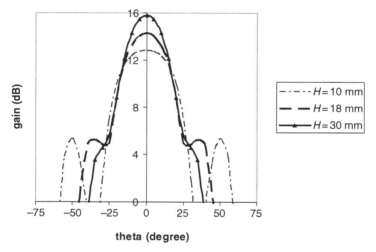

Figure 3.20. Simulated radiation patterns for different heights (*H*) of a tapered PTFE polyrod at phi = 0 (© 2006 IEEE [37]).

To measure the reflection coefficient, an Anritsu VP connector was used as an interface between the antenna and the network analyzer used to make the measurement [37]. The measured S_{11} (square of reflection coefficient) of the antenna is plotted in Figure 3.21, which also shows the response for the patch antenna on its own and where we see the characteristic, rather notch-like response typical of this antenna type. Here it is quite evident that when the tapered polyrod is added on top of the patch, so that the patch acts as the feed and the polyrod as the driven element, the input match of the combined antenna is very much better than that of the patch alone.

Table 3.1 shows how a slightly higher gain can be achieved by tuning the height of the waveguide and also by adjusting that of the tapered polyrod while keeping the

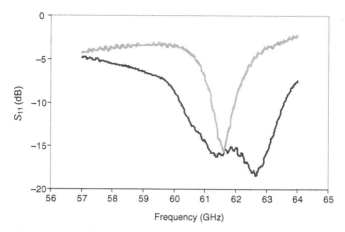

Figure 3.21. Measured S_{11} of the patch antenna with a tapered polyrod (lower black line) and without a polyrod (upper gray line) (© 2006 IEEE [37]).

TABLE 3.1. Calculated Antenna Gain at Different Waveguide Height and Tapered Polyrod Height (The Height of Cylindrical Polyrod = 3 mm) (© 2006 IEEE [37])

Waveguide Height	Tapered Polyrod Height		
	24 mm	27 mm	30 mm
3.5 mm	15.22 dBi	15.46 dBi	15.74 dBi
7 mm	16.19 dBi	16.59 dBi	16.74 dBi

height of the cylindrical section fixed at 3 mm. When the height of the waveguide increased from 3.5 to 7.0 mm, the gain increased by approximately 1 dB. If the height of the waveguide is increased further, the gain is not further improved because the metallic waveguide has reached resonance and so reduces the radiation from the tapered polyrod.

Measured E-plane radiation patterns of the patch-fed tapered polyrod are shown for two frequencies in Figure 3.22. Again, the pattern for the patch alone is also included here, where we see the characteristic broad beamwidth.

Figure 3.23 then compares simulation and measurement results of the maximum gain of the patch-fed polyrod as a function of frequency. The gain of the combined antenna sees an increase of up to 15 dB, in the frequency band of 59–65 GHz, compared to the patch feed alone. Also, the gain flatness of the patch-fed polyrod, across this band, is in the region of 2 dB. This characteristic is especially useful for an extremely high-rate (beyond gigabit per second) communications system where a complex equalizer might be saved.

When the shape of a polyrod is designed asymmetrically as shown in Figure 3.18c and the cut angle is set to 60°, the RHCP is present, while the LHCP is suppressed.

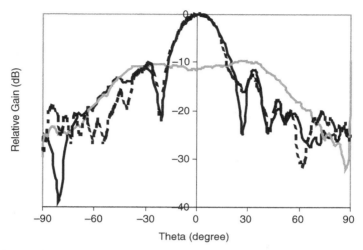

Figure 3.22. Measured *E*-plane radiation patterns of patch-fed tapered polyrod at 61 GHz (solid black line) and 63 GHz (dotted black line) and pattern of patch alone at 61 GHz (gray line). (Curves are normalized to the maximum of the polyrod pattern at 61 GHz.) (© 2006 IEEE [37]).

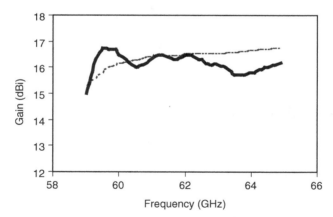

Figure 3.23. Measured (solid line) and simulated (dotted line) maximum gain of the patch-fed polyrod antenna versus frequency (© 2006 IEEE [37]).

The axial ratio for the patch-fed polyrod was measured from 59 to 63 GHz as shown in Figure 3.24. The cross polarization is determined by the cutoff angle and the design of the circular-polarized patch. By tuning these two parameters, the cross polarization level can be tailored to fit various applications.

The beam direction can be adjusted to different directions by tilting the axis of the polyrod as shown in Figure 3.19. When considering the effect of integrating this antenna

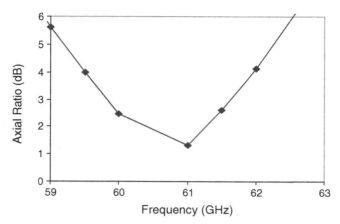

Figure 3.24. Measured axial ratio of the patch-fed polyrod antenna with the geometry of Figure 3.18c (© 2006 IEEE [37]).

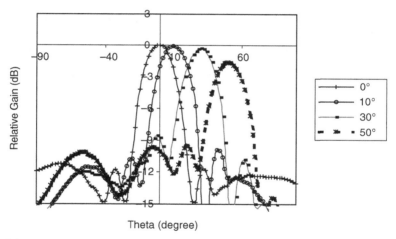

Figure 3.25. Measured *E*-plane radiation patterns with beam directions 0°, 10°, 30°, and 50° (normalized to the 0° case) (© 2006 IEEE [37]).

with consumer devices, the ground plane should be electrically large; a size of the order of 20 × 20 wavelengths has been shown to be effective.

The measurement results in Figure 3.25 show beams at scan angles of 0°, 10°, 30°, and 50°, which were brought about by tilting the polyrod to these angles. (The beam scan angle is equal to the tilt angle of the polyrod.) For the case at a tilt angle of 50°, the asymmetrical effect of the radiation to the ground (*x-y*) plane becomes significant, and so the sidelobe level increases and the gain drops off.

To close this section on a practical note, some low constant polymers are not very stiff mechanically and this can be an impediment to obtaining good dimensional tolerances for polyrods. PTFE scores badly in this respect, and other materials like polyethylene or polypropylene, while good, low-loss dielectrics, suffer similarly. A way to improve matters is to chill the material to a low temperature in order to increase its hardness prior machining. Cross-linked polystyrene (where $\varepsilon_r \approx 2.5$) is a harder material at room temperature, has better machinability, and so could be easier to apply to polyrods, compared to PTFE, if the slight increase in permittivity can be accommodated during the design.

3.4 ARRAY OF POLYRODS

To increase the gain of tapered polyrod antennas above the 18- to 20-dB limit, *arrays* of these antennas can be used. Since the individual array elements already have a substantial directivity, element spacing can be chosen to be comparatively large so that interelement coupling is minimized. According to Zucker [13], the spacing b should be in the range

$$0.5\sqrt{\frac{L}{\lambda_0}} \le \sqrt{\frac{b}{\lambda_0}} \le \sqrt{\frac{L}{\lambda_0}}, \tag{3.13}$$

where L is the length of the polyrod.

The lower bound formulates the condition that interelement coupling should be negligible. The upper bound is determined by the requirement that the first grating lobe should be substantially reduced by multiplication with the element pattern. Ideally, the first minimum of the element pattern should coincide with the first grating lobe of the array pattern. For scanned arrays, the element spacing b must be made smaller, in correspondence with the desired scan range. Under these circumstances, the array gain of the main beam of a linear array of n equally spaced elements is approximated by

$$G_{\text{array}} \approx nG_{\text{element}}. \tag{3.14}$$

The performance of a polyrod array depends on the geometry and dimensions of the polyrod elements, the spacing between elements, the number of elements, the mode of operation, and the feed arrangement. Various polyrod geometries discussed in the previous sections offer a range of choices of polyrod elements, and a number of structures can be used to feed polyrod arrays [38]. The polyrods and feed arrangements, which have been used for generating linear and circular polarization, are summarized in Table 3.2, and three commonly used feed arrangements, that is, slot coupling, microstrip coupling, and probe coupling, are illustrated in Figure 3.26a–c.

In common with other antenna arrays, the radiation pattern of a polyrod array is determined by the radiation pattern of a single polyrod element and the array factor (*AF*). In general, the radiated electric field in the far-field region of a polyrod array can be written in the form

$$E = n \times E_0 \times AF, \tag{3.15}$$

TABLE 3.2. Choice of Polyrod Shape and Feeding Method

Polyrod Array (Polarization)	Feeding Method
• Cross-shaped polyrods (circular)	• Probe excitation with microstrip feed
• Cylindrical polyrods (linear or circular)	• Microstrip feed
• Geometrically modified rectangular polyrods (circular)	• Slot coupling with microstrip feed
	• Microstrip branch line feed with slot coupling
• Hemispherical polyrods (linear)	
• Multisegment rectangular polyrods (linear)	• Dielectric waveguide
• Rectangular polyrods (linear)	• Dielectric image guide
	• Coplanar waveguide feed

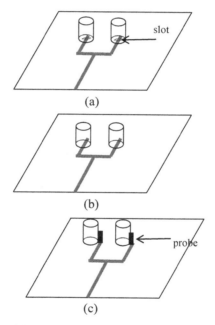

Figure 3.26. Illustration of (a) slot, (b) microstrip, and (c) probe coupling arrangements of two-polyrod arrays.

where n is the number of elements of the array; E_0 is the electric field at the same field point produced by a single polyrod positioned at the center of the array; and AF is the array factor. The array factor depends on the geometric structure of the array, the phase difference of currents between elements, and the operating frequency.

The array factor for a linear E-plane array with in-phase feed currents and an array axis in the x-direction, as shown in Figure 3.27a, can be expressed as

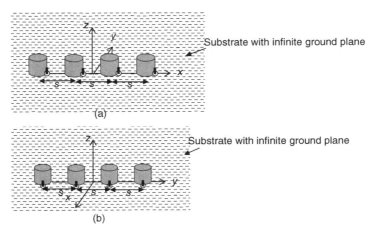

Figure 3.27. (a) E-plane and (b) H-pane arrays with an infinite ground plane.

$$AF_E = \frac{\sin[n\Psi_E/2]}{n\sin[\Psi_E/2]}, \qquad (3.16)$$

where $\Psi_E = k_0 s \sin\theta \cos\phi$, $k_0 = 2\pi/\lambda_0$ is the wave number in free space, λ_0 is the free-space wavelength, and s is the spacing between elements.

The array factor for a linear H-plane array with an array axis in the y-direction, as shown in Figure 3.27b, is

$$AF_H = \frac{\sin[n\Psi_H/2]}{n\sin[\Psi_H/2]}, \qquad (3.17)$$

where $\Psi_H = k_0 s \sin\theta \sin\phi$.

For different element numbers $n = 2$, 4, and 8 and spacing $s = 0.5\lambda_0$, the array factor on the $\phi = 0$ plane for the E-plane array (Fig. 3.27a) with an infinite ground plane is shown in Figure 3.28. Because of the ground plane, only positive angles apply. The pattern of the array factor on the $\phi = 0$ plane is directional in $\theta = 0$ direction for $s < \lambda_0$. The directivity increases with the number of elements. However, when $s > 0.5\lambda_0$, the sidelobes along the $\theta = \pi/2$ direction start to develop, and the level of sidelobes increases significantly when $s > 0.6\lambda_0$. Hence, the spacing of $0.5\lambda_0$ is commonly used in the design and study of polyrod arrays.

For polyrods operating at the fundamental mode of the geometry with broadside radiation, the E-plane array with a typical spacing of $0.5\lambda_0$ has higher directivity than the H-plane array.

For a planar array of $(m \times n)$ elements with in-phase feed currents, the array factor can be written as

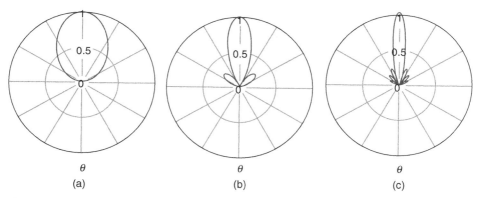

Figure 3.28. Normalized magnitude of array factor, *AF* for infinite ground plane and $\phi = 0$ cut, of *E*-plane arrays with (a) $n = 2$, (b) $n = 4$, and (c) $n = 8$. Spacing $s = 0.5\ \lambda_0$.

$$AF = \frac{1}{mn} \frac{\sin[m(k_0 s_x \sin\theta\cos\phi)/2]}{\sin[(k_0 s_x \sin\theta\cos\phi)/2]} \frac{\sin[n(k_0 s_y \sin\theta\sin\phi)/2]}{\sin[(k_0 s_y \sin\theta\sin\phi)/2]}, \qquad (3.18)$$

where s_x and s_y are the spacings between elements in the *x*- and *y*-directions, respectively.

The radiation of polyrods is operating at the fundamental mode, namely, the $HE_{11\delta}$ mode for the cylindrical rod and the $TE_{11\delta}$ mode for the rectangular rod. Their arrays can equally be modeled as arrays of magnetic dipoles. For the same number of *n* or *m* and *n*, the *E*-plane and planar arrays are more directional in the broadside direction.

The formation of a directional polyrod array requires that the polyrod elements be positioned with a close spacing between elements. However, the proximity of the elements results in mutual coupling between elements due to the electromagnetic interaction between elements. The mutual coupling may be significant so that it substantially affects the performance of the array including radiation pattern, directivity, operating frequency, and bandwidth. It also complicates the design of the matching network.

The level of mutual coupling depends not only on the spacing between elements but also on the structure, dimensions, and dielectric constant of the polyrod elements and the mode of operation. The simulation and experimental studies of slot-coupled cylindrical polyrods operating at the fundamental $HE_{11\delta}$ mode with a diameter of 5.12 mm, height of 3.02 mm, and a relative dielectric constant $\varepsilon_r = 10$ by Guo et al. [39] show that for spacing $s > 0.35\ \lambda_0$, the level of mutual coupling, that is, $|S_{21}|$, between polyrod elements for both *E*-plane and *H*-plane arrays, is less than -10 dB at 16 GHz. The simulated and measured results are shown in Figure 3.29. However, the mutual coupling in the *H*-plane array is stronger than that in the *E*-plane array when $s \leq 0.5\ \lambda_0$.

When the spacing, is large (e.g., $s > 0.6\ \lambda_0$), the mutual coupling in the *E*-plane array becomes stronger than that in the *H*-plane array.

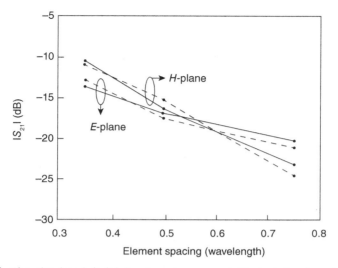

Figure 3.29. The simulated (solid lines) and measured (dashed lines) mutual coupling between two slot-coupled cylindrical polyrods with diameter 5.12 mm, height 3.02 mm and $\varepsilon_r = 10$, in E-plane and H-plane array configurations (© 1999 IEEE [39]).

When s is half-wavelength, the level of mutual coupling for both E-plane and H-plane arrays is less than -15 dB. Loos and Antar [40] show that the level of mutual coupling for both E-plane and H-plane arrays operating at 7 GHz is approximately -15 dB at $s = 0.5\ \lambda_0$, and the E-plane array has a stronger coupling when $s > 0.5\ \lambda_0$. The mutual coupling could, however, reach -9 dB for $s \approx 0.3\ \lambda_0$.

Slot coupling with a microstrip feed is one of the most commonly used and simple methods for the excitation of polyrod arrays. It makes use of the electromagnetic coupling from a microstrip line to the polyrod through a nonradiating slot on the ground plane, as illustrated in Figure 3.30a for a cylindrical polyrod array. Both E-plane and H-plane linear arrays can be fed using slot coupling and microstrip corporate feed. Using slot coupling, the slot is usually positioned at a distance of $\lambda g/4$ away from the open end of the microstrip line, where λg is the guided wavelength of the microstrip guiding structure. The width and length of the slot need to be appropriately chosen so that good matching to a 50-Ω microstrip line is obtained for each polyrod element.

Four-element linear E-plane and H-plane cylindrical polyrod arrays using the slot coupling were demonstrated by Drossos et al. [41]. The fundamental $HE_{11\delta}$ mode of the polyrods was used to produce broadside radiation. Microstrip corporate feeds were also used for both E-plane and H-plane configurations, as shown in Figure 3.30a,b, respectively.

Polyrod elements can also be fed in series using an open-ended microstrip line as shown in Figure 3.31, which was first demonstrated by Petosa et al. [10, 42]. The elements are separated by approximately one guided wavelength, and they are placed along the microstrip line on one side with different coupling coefficients. With the use of 10 rectangular elements of dimensions $a = 9$ mm, $b = 7.5$ mm, and $d = 3.1$ mm and

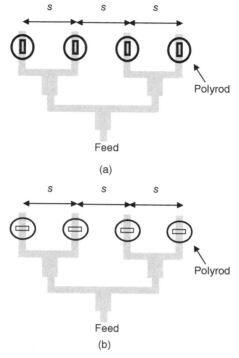

Figure 3.30. Four-element slot-coupled cylindrical polyrod array in (a) *E*-plane and (b) *H*-plane configurations.

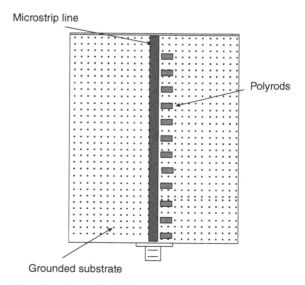

Figure 3.31. Top view of a microstrip-fed series rectangular polyrod array (© 1998 IEEE [10]).

$\varepsilon_r = 10.8$ on a ground plane of 30×30 cm, a gain of 13.2 dBi was obtained at 8 GHz with a cross-polarization level of 20 dB below the copolarized peak value and sidelobe levels more than 17.5 dB below the peak value.

The array has a narrow impedance bandwidth (~2%) as a result of the strong frequency dependence of the loaded microstrip line structure. However, Petosa et al. [43] showed that bandwidth could be improved significantly using a paired polyrod arrangement. In this structure, polyrods in each pair are separated slightly, with each element in the pair located on alternate sides of the microstrip line so as to reduce the mutual coupling between elements. A 10-dB return loss bandwidth of 18% was obtained for an array of 16 rectangular polyrod elements. The gain bandwidth is also improved using the paired structure.

At millimeter wave frequencies, the losses on the microstrip line in the series array become very significant. The microstrip line may then be replaced by a dielectric waveguide or an image guide, as illustrated by Birand and Gelsthorpe [44]. The dielectric waveguide coupled traveling wave polyrod array studied in Reference 44 operates in the vicinity of 32.5 GHz. The array has a beamwidth of 8.5°, which could scan approximately 20° from the broadside direction at 35.5 GHz.

Compared with linear polyrod arrays of the same number of elements with microstrip corporate feed and slot or probe coupling, the gain of the microstrip-fed series array is generally significantly lower. The main beam of the series array also has a strong frequency dependence, which precludes its use for wideband fixed-beam applications [10].

To generate circular polarization from an array, circular polarized (CP) elements can of course be used. An alternative approach is to use phase-alternate linear polarized (LP) feeds—a technique often used in arrays of at least four elements where adjacent radiators of both horizontal and vertical relative polarization are fed with a quadrature differential phase shift. A 2×2 polyrod array configured according to this recipe is shown in Figure 3.32. This is also a low-profile antenna employing quite short polyrods. A version has also been developed with integrated amplifiers, digital phase shifters, and filters providing electronic beam-steering capability. The goal was also to obtain a low-profile, active phased array with beam-steering capability. The relative orientation of each patch is rotated 90° clockwise around the center point of array [45]. Thus, the feed to each patch has a 90° angular orientation as shown in Figure 3.32. With this spacing, the fields generated from the four feeds are orthogonal to each other. Additionally, the four feeds are required to be fed 90° out of time phase for achieving TM_{11} mode circular polarization. The phase of each patch, therefore, is 0°, 90°, 180°, and 270°, respectively, in a clockwise direction [45]. For LHCP, the electric field vector will rotate in the opposite direction. The phase settings for both RHCP and LHCP are illustrated in Figure 3.33.

Studies show that polyrod arrays have various advantages over single polyrod elements including directional radiation patterns, increased gain, and often wider bandwidths. They also generally have low cross-polarization levels. These arrays are competitive with other types of antenna arrays, in terms their performance and fabrication process. They will *predominately* find applications in point-to-point, point-to-multipoint, and satellite communications systems, and also in dynamic radar systems.

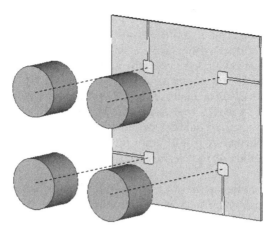

Figure 3.32. Configuration of a 2 × 2 polyrod array (polyrods are directly located on top of patches).

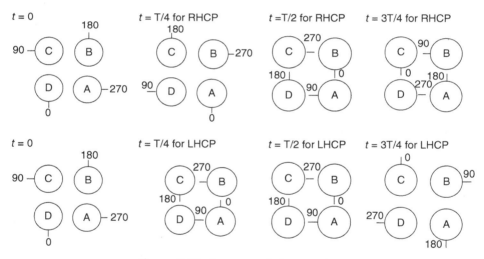

Figure 3.33. Summary of phase settings.

For simplicity and convenience, microstrip lines are usually used in the feed network of polyrod arrays in the lower gigahertz range of microwave frequencies. High operation efficiency can be obtained for small arrays. At millimeter wave frequencies, the efficiency becomes significantly lower due to the losses on the microstrip lines. Therefore, it would be beneficial to use low-loss dielectric guides to construct the feed network for polyrod arrays. The superior performance of polyrod arrays over microstrip antenna arrays at millimeter wave frequencies should then be revealed.

The ultimate performance of polyrod arrays may be affected by a number of factors:

- nonuniformity of dielectric materials between elements,
- fabrication tolerance of polyrod elements and the feed structure,
- mutual coupling between elements and within the feed structure,
- loss in the feed structure,
- finite-size ground plane,
- polyrod mounting technology,
- air gaps, and
- the presence of protection enclosure.

These effects need to be considered in the design of polyrod arrays. Nonuniformity, fabrication tolerance, and air gaps should be minimized in the manufacturing process so that polyrod arrays with desirable performance or least postfabrication adjustments may be manufactured.

3.5 MULTIBEAM POLYROD ARRAY

In a multipath environment, antennas with a narrow beam can be used to reduce the number of multipath signals and therefore to minimize the root-mean-square delay spreads. A narrow beam antenna has a high directivity that directs or confines the power or reception in a given direction and thus extends the communication range. Furthermore, the gain of the antennas can partly reduce the required gain of millimeter wave power amplifiers by supplying more captured power to the output terminals of the antenna, so the power consumption in microwave and millimeter wave circuitry is potentially reduced.

Conventional millimeter wave links can be classified based on whether or not an uninterrupted line-of-sight (LOS) is established between the transmitter and the receiver. For indoor applications, non-line-of-sight (NLOS) scenarios, also called diffuse links, are very common [47]. Conventional millimeter wave communications systems, whether LOS or NLOS, mostly employ a single antenna. This section looks one of the most promising ways to improve the performance, that is, using a multibeam directional array, which utilizes multiple beams that are pointed in different directions [47]. Such an angle-diversity antenna array can have an overall high directivity and more information capacity.

The concept also offers the possibility of reducing the effects of cochannel interference and multipath distortion because the unwanted signals are angularly filtered out by the individual narrow beams. The multibeam antenna array can be implemented using multiple polyrods which are oriented in different directions. A conventional polyrod antenna is fed by waveguides [1], an arrangement that is too bulky for the particular array structure under consideration here. The more attractive approach for the polyrod configuration uses a patch-fed method [36] to make a polyrod array with the advantage that it can be integrated with a variety of planar circuits.

The geometry of the antenna array is shown in Figure 3.34. The central polyrod antenna no. 1 is in an upright direction, which is perpendicular to the plane of the

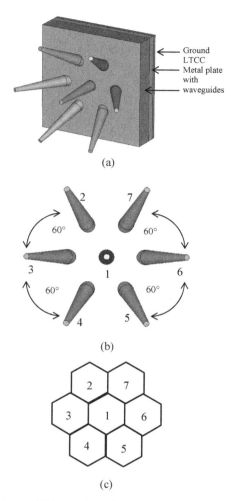

Figure 3.34. (a) Geometry and (b) top view of a seven-polyrod antenna. (c) Radiation coverage in hexagonal configuration. LTCC, low temperature co-fired ceramic. (© 2006 IEEE [46].)

patches, while the other polyrods (nos. 2, 3, 4, 5, 6, and 7) are tilted with a polar angle (θ) of 40° relative to the surface normal, toward the plane of the patches. The tilted antennas are rotated with respect to the central polyrod and have an azimuthal angular spacing ($\Delta\Phi$) of 60°. Seven polyrod antennas are fitted into a metal plate with corresponding feeds. Each polyrod has a different angular radiation coverage. When the antennas cover a given spatial area, each polyrod covers a nominally nonoverlapping cell in a similar arrangement to a cellular system, as shown in Figure 3.34c.

The center frequency of the antenna described here is designed for 60 GHz. As shown in Figure 3.35, the polyrods are made of Teflon® (PTFE) with a 3-mm-diameter

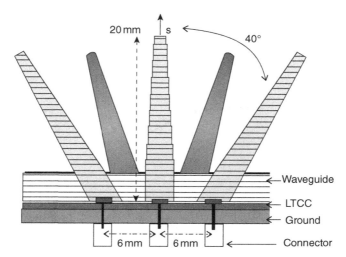

Figure 3.35. Side view of the polyrod antennas (© 2006 IEEE [46]).

cylindrical base. The upper part of polyrod is tapered linearly to a terminal aperture with a 0.6-mm diameter to reduce minor lobes in the radiation pattern. The total height of the central polyrod antenna is 20 mm to ensure a high antenna gain. The diameter of the central polyrod (no. 1) antenna is designed to be 3 mm. Each polyrod antenna is designed to have a 40° half-power beamwidth. The whole radiation zone of the seven polyrod antennas, therefore, covers a polar angle of 60° approximately with respect to the z-axis, which is the axis perpendicular to the plane of the patches.

The fields at the polyrod surfaces are derived using equivalent electric and magnetic current sheets, and the radiation field is simulated from these currents. The relative electric field pattern E as a function of the polar angle θ from the normal axis is derived by the following formula [36]:

$$E(\theta) = (\sin \Phi)/\Phi, \qquad (3.19)$$

where $\Phi = H_\lambda \pi (\cos \theta - 1) - 0.5 \pi$, and H_λ is the height of the polyrod in free-space wavelengths.

The tapered polyrod can be treated as an impedance transformer and it reduces the reflection that would be caused by an abrupt discontinuity [28]. The polyrods are fed by patches on low-temperature cofired ceramics substrates [37, 48]. The patch-fed method can adapt polyrod antennas to most planar circuits. It firstly saves feeding space and also brings more design flexibility to the array structure. Additionally, it increases the directivity and bandwidth of conventional patch antennas alone.

The patches are either circular or elliptical in shape to match the base of the polyrods. The patches are individually energized by probes connected to coaxial connectors. The probe feed can provide smaller sidelobes in comparison with microstrip line feeds because the coupling between antennas and the feeding lines is limited by the ground plane.

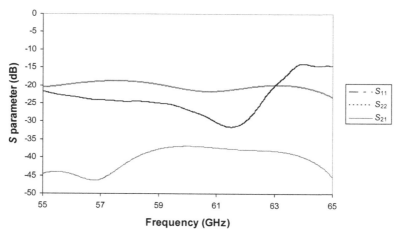

Figure 3.36. Measured S parameters for upright polyrod—no. 1 (S_{11}), one of the tilted polyrods—no. 2 (S_{22}) and their mutual coupling (S_{21}) (© 2006 IEEE [46]).

The performance of the above antennas was first simulated using CST Microwave Studio, which is a commercial simulator package based on the finite integration time-domain method. The spacing between adjacent polyrods was set to 6 mm to allow enough space between the test connectors, which would be needed for the later practical measurements. As can be seen in Figure 3.36, the antenna can be operated from 55 to 65 GHz with a return loss of about 23 dB. Polyrod antennas show a broad impedance bandwidth and are suitable for wireless personal area networks such as IEEE 802.15.3c related applications [49]. The coupling between adjacent polyrods (e.g., nos. 1 and 2, nos. 3 and 2) was measured to be approximately −40 dB. Because of the symmetric configuration, any two pairs of polyrods with the same relative spacing have similar coupling coefficients. Appropriate grounding of the conducting plate is important to reduce coupling due to surface-wave propagation.

The 60-GHz radiation pattern is shown in a Cartesian plot in Figure 3.37. It depicts the 60-GHz radiation pattern of the upright polyrod (no. 1) and a tilted polyrod (no. 3) at $\theta = 0°$ plane. The main beam is in the direction of $\theta = 0°$ and −40° for the upright polyrod and the tilted polyrod, respectively. It is shown that the maximum gain of the polyrod (no. 3) is radiated at the same direction as the physical axis of the polyrod. The upright polyrod has a half-power beamwidth at $\theta = -20°$, while the tilted polyrod has a half-power beamwidth between $\theta = -14°$ and −53°. Because of asymmetric shape of the tilted polyrod, one side has a longer height than the other. The side with long height can radiate and receive more energy than the short side. Thus, its radiation pattern is asymmetric, as shown in Figure 3.37. The curves are normalized to the pattern of the upright polyrod to show the relative power. For the tilted polyrod, there is a sidelobe at 13° with the level of about −8.5 dB.

The frequency response was measured by a comparison method with a V-band standard horn antenna. As can be seen in Figure 3.38, it was observed that an average

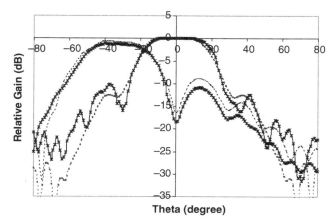

Figure 3.37. Simulated (–) and measured (-*-) radiation pattern at phi = 0° plane as a function of elevation angle at 60 GHz when main beam is at the direction of theta = 0° and –40°, respectively. These curves are normalized to that of the central polyrod no. 1 (© 2006 IEEE [46]).

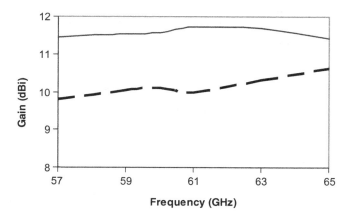

Figure 3.38. Maximum gain of upright polyrod (solid line) and tilted polyrod (dashed line) between 57 and 65 GHz (© 2006 IEEE [46]).

antenna gain of about 11.5 and 9.8 dBi was measured between 57 and 65 GHz for the upright polyrod and the tilted polyrods, respectively. The 3-dB bandwidth of both the upright antenna and the tilted antenna was approximately 19% of the center frequency, which is higher than the 11% bandwidth reported in Reference [50]. By comparing the measured antenna gain and the directivity, the radiation efficiency of the implemented prototype was estimated to be 80% and 73% for the upright and the tilted polyrods, respectively, while the aperture efficiency was 74% and 57%, respectively. Note that the main beam is fairly circularly symmetric in its half-power beamwidth region, and therefore the interelement angle spacing ensures a 3-dB radiation pattern overlapping in any principal cut in the azimuthal plane.

REFERENCES

[1] S. Kobayashi, R. Miptra, and R. Lampe, "Dielectric Tapered Rod Antennas for Millimeter-Wave Applications," IEEE Trans. Antennas Propag. Vol. AP-30, No. 1, 1982, pp. 54–58.

[2] K.-M. Luk and K.-W. Leung, Dielectric Resonators, Artech House, Norwood, MA, 1986, p. 197.

[3] D. Hondros, "Ueber elektromagnetische Drahtwelle," Annalen der Physik Vol. 30, 1909, pp. 905–949.

[4] H. Zahn, "Ueber den Nachweis elektromagnetischer Wellen an dielektrischen Draehten," Annalen der Physik Vol. 37, 1916, pp. 907–933.

[5] S. A. Long, M. W. McAllister, and L. C. Shen, "The Resonant Cylindrical Dielectric Cavity Antenna," IEEE Trans. Antennas Propag. Vol. 31, 1983, pp. 406–412.

[6] R. K. Mongia and A. Ittipiboon, "Theoretical and Experimental Investigations on Rectangular Dielectric Resonator Antennas," IEEE Trans. Antennas Propag. Vol. 45, 1997, pp. 1348–1356.

[7] R. K. Mongia, A. Ittipiboon, P. Bhartia, and M. Cuhaci, "Electricmonopole Antenna Using a Dielectric Ring Resonator," Electron. Lett. Vol. 29, 1993, pp. 1530–1531.

[8] A. W. Glisson, D. Kajfez, and J. James, "Evaluation of Modes in Dielectric Resonators Using a Surface Integral Equation Formulation," IEEE Trans. Microwave Theory Tech. Vol. MTT-31, 1983, pp. 1023–1029.

[9] R. K. Mongia and P. Bhartia, "Dielectric Resonator Antennas—A Review and General Design Relations for Resonant Frequency and Bandwidth," Int. J. Microwave and Millimeter-Wave Computer-Aided Eng. Vol. 4, No. 3, 1994, pp. 230–247.

[10] A. Petosa, A. Ittipibon, Y. M. M. Antar, D. Roscoe, and M. Cuhaci, "Recent Advances in Dielectricresonator Dielectric Resonator Antenna Technology," IEEE Antenn. Propag. Mag. Vol. 40, No. 3, 1998, pp. 35–48.

[11] A. A. Kishk, A. W. Glisson, and D. Kajfez, "Computed Resonant Frequency and Far Fields of Isolated Dielectric Discs," IEEE Antennas and Propagation Society International Symposium Digest, Vol. 1, pp. 408–411, 1993.

[12] L. B. Felsen, "Radiation from a Tapered Surface Wave Antenna," IRE Trans. Antennas Propag. Vol. AP-8, 1960, pp. 577–586.

[13] F. J. Zucker, Antenna Engineering Handbook—Surface and Leaky-Wave Antenna, 3rd ed., McGraw-Hill, New York, 1992.

[14] R. E. Collin and F. J. Zucker, Antenna Theory, McGraw-Hill, New York, 1969.

[15] J. B. Andersen, Metallic and Dielectric Antennas, Polyteknisk Forlag, Odense, Denmark, 1971.

[16] J. R. James, "Engineering Approach to the Design of Tapered Dielectric-Rod and Horn Antennas," Radio Electron. Eng. Vol. 42, No. 6, 1972, pp. 251–259.

[17] T. Takano and Y. Yamada, "The Relation between the Structure and the Characteristics of a Dielectric Focused Horn," Trans. IECE Vol. J60-B, No. 8, 1977, pp. 395–593.

[18] C. Yao and S. E. Schwarz, "Monolithic Integration of a Dielectric Millimeter-Wave Antenna and Mixer Diode: An Embryonic Millimeter-Wave IC," IEEE Trans. Microwave Theory Tech. Vol. MTT-30, 1982, pp. 1241–1247.

[19] R. Chatterjee, Dielectric and Dielectric-Loaded Antennas, Research Studies Press, Baldock, U.K., 1985.

[20] F. Schwering and A. A. Oliner, "Millimeter-Wave Antennas," in Antenna Handbook, Y. T. Lo and S. W. Lee, eds., Van Nostrand Reinhold, New York, 1988, Chapter 17.

[21] M. Koshiba, Optical Waveguide Analysis, McGraw-Hill, New York, 1990, Chapter 5.

[22] C. Vassallo, "1993–1995 Optical Mode Solvers," Optic. Quantum Electron. Vol. 29, 1997, pp. 95–114.

[23] D. Yevick and W. Bardyszewski, "Correspondence of Variation Finite Difference (Relaxation) and Imaginary-Distance Propagation Methods for Modal Analysis," Opt. Lett. Vol. 17, No. 5, 1992, pp. 329–330.

[24] S. Xiao, R. Vahldieck, and H. Jin, "Full-Wave Analysis of Guided Wave Structure Using a Novel 2-D FDTD," IEEE Microw. Guid. Wave Lett. Vol. 2, 1992, pp. 165–167.

[25] A. Asi and L. Shafai, "Dispersion Analysis of Anisotropic Inhomogeneous Waveguides Using Compact 2D-FDTD," Electron. Lett. Vol. 28, 1992, pp. 1451–1452.

[26] S. M. Lee, "Finite-Difference Vectorial-Beam-Propagation Method Using Yee's Discretization Scheme for Modal Fields," J. Opt. Soc. Amer. A, Opt. Image Sci. Vol. 13, No. 7, 1996, pp. 1369–1377.

[27] J. Yamauchi, N. Morohashi, and H. Nakano, "Rib Waveguide Analysis by the Imaginary-Distance Beam-Propagation Method Based on Yee's Mesh," Optic. Quantum Electron. Vol. 30, 1998, pp. 397–401.

[28] T. Ando, J. Yamauchi, and H. Nakano, "Demonstration of the Discontinuity-Radiation Concept for a Dielectric Rod Antenna," in Proceedings of IEEE Antennas and Propagation Society International Symposium, Jul. 16–21, vol. 2, pp. 856–859, 2000.

[29] S. T. Chu, W. Huang, and S. K. Chaudhuri, "Simulation and Analysis of Waveguide Based Optical Integrated Circuits," Comput. Phys. Communicat. Vol. 68, 1991, pp. 451–484.

[30] A. Taflove and S. C. Hagness, Computational Electrodynamics, the Finite-Difference Time-Domain Method, 2nd ed., Artech House, Norwood, MA, 2000.

[31] O. M. Ramahi, "The Concurrent Complementary Operators Method for FDTD Mesh Truncation," IEEE Trans. Antennas Propag. Vol. 46, 1998, pp. 1475–1482.

[32] D. B. Rutledge et al., "Antennas and Waveguides for Far-infrared Integrated Circuits," IEEE J. Quantum Electron. Vol. QE-16, May 1980, pp. 508–516.

[33] T. N. Trinh, J. A. Malberk, and R. Mlttra, "A Metal-to-Dielectric Waveguide Transition with Application to Millimeter-Wave Integrated Circuits," IEEE MTT-S Int. Microwave Symp. 1980, pp. 205–207.

[34] J. A. G. Malherbe, T. N. Trinh, and R. Mittra, "Transition from Metal to Dielectric Waveguide," Microwave J. Vol. 23, November 1980, pp. 71–74.

[35] S. M. Hanham, T. S. Bird, A. D. Hellicar, and R. A. Minasian, "Optimized Dielectric Rod Antennas for Terahertz Applications," 34th International Conference on Infrared, Millimeter, and Terahertz Waves, 2009. IRMMW-THz 2009, pp.1–2, 2009.

[36] J. Kraus and R. Marhefka, Antennas for All Applications, 3rd ed., McGraw Hill, New York, 2002.

[37] K. Huang and Z. Wang, "V-Band Patch-Fed Rod Antennas for High Data-Rate Wireless Communications," IEEE Trans. Antennas Propag. Vol. 54, No. 1, January 2006, pp. 297–300.

[38] A. Petosa, R. K. Mongia, A. Ittipiboon, and J. S. Wight, "Experimental Investigation on Feed Structures for Linear Arrays of Dielectric Resonator Antennas," Proceedings of IEEE AP-S Conference, California, USA, pp. 1982–1985, 1995.

[39] Y. X. Guo, K. M. Luk, and K. W. Leung, "Mutual Coupling between Millimeter-Wave Dielectric Resonator Antennas," IEEE Trans. Microwave Theory Tech. Vol. 47, No. 11, 1999, pp. 2164–2166.

[40] G. D. Loos and Y. M. M. Antar, "A New Aperture-Coupled Rectangular Dielectric Resonator Antenna Array," Microwave Opt. Tech. Lett. Vol. 7, No. 14, 1994, pp. 642–644.

[41] G. Drossos, Z. Wu, and L. E. Davis, "Aperture-Coupled Cylindrical Dielectric Resonator Antennas Forming Four-Element Linear Array," Microwave Opt. Tech. Lett. Vol. 20, No. 2, 1999, pp. 151–153.

[42] A. Petosa, R. K. Mongia, A. Ittipiboon, and J. S. Wight, "Design of Microstrip-Fed Series Array of Dielectric Resonator Antennas," Electron. Lett. Vol. 31, No. 16, 1995, pp. 1306–1307.

[43] A. Petosa, A. Ittipiboon, M. Cuhaci, and R. Larose, "Bandwidth Improvement for a Microstrip-Fed Series Array of Dielectric Resonator Antennas," Electron. Lett. Vol. 32, No. 7, 1996, pp. 608–609.

[44] M. T. Birand and R. V. Gelsthorpe, "Experimental Millimetric Array Using Dielectric Radiators Fed by Means of Dielectric Waveguides," Electronics Letters Vol. 17, No. 18, 1981, pp. 633–635.

[45] K. Huang, S. Koch, and M. Uno, "Circular Polarized Array Antenna," European Patent EP1564843, 2005.

[46] K. Huang and D. J. Edwards, "60 GHz Multi-beam Antenna Array for Gigabit Wireless Communication Networks," IEEE Trans. Antennas Propag. Vol. 54, No. 12, 2006, pp. 3912–3914.

[47] M. Uno, Z. Wang, V. Wullich, and K. Huang, "Communication System and Method," Patent No. US2006116092, EP1659813, JP2006148928, 2006.

[48] K. Huang and Z. Wang, "Dielectric Rod Antenna and Method for Operating the Antenna," Patent WO2006097145, EP1703590, September 21 2006.

[49] IEEE 802.15.3c Standard, http://www.ieee.org/

[50] T. Ando, J. Yamauchi, and H. Nakano, "Rectangular Dielectric-Rod by Metallic Waveguide," IEE Proc. Microwave, Antennas Propag. Vol. 149, No. 2, 2002, pp. 92–97.

4

MILLIMETER WAVE
LENS ANTENNAS

Kao-Cheng Huang

Lens antennas are more attractive in the millimeter wave region than at lower frequencies because their dielectric weight is significantly reduced. For a given electrical performance, the volume and weight of a lens decrease with λ_0^3 and are usually very reasonable at millimeter wavelengths. Because of the reduction in dielectric weight, *spherically symmetric lenses* also become practical in the millimeter wave region [1]. Luneburg lenses which focus an incident plane wave into a single point on the lens surface would provide excellent performance, but the required gradation of the refractive index profile of these lenses would probably be difficult to realize in the millimeter wave region. Moreover, much simpler lenses, for example, homogeneous spheres, provide already good performance, and two-layer lenses consisting of a homogeneous spherical core surrounded by a concentric shell of a different refractive index can be sufficiently well corrected to satisfy almost all practical requirements [1].

In addition to their simple configuration and their good electrical performance, such spherically symmetric lenses permit wide-angle beam scanning without pattern degradation. If the antennas are operated with an array of identical feed horns, beam scanning can be implemented digitally by electronic switching from feed horn to feed horn.

Modern Lens Antennas for Communications Engineering, First Edition. John Thornton and Kao-Cheng Huang.
© 2013 Institute of Electrical and Electronics Engineers. Published 2013 by John Wiley & Sons, Inc.

The lens acts as a concentrator, gathering energy over an area and concentrating it to a point. Tolerance requirements of lens are less stringent than for reflector antennas since lenses are usually made from materials of relatively low refractive index to minimize reflection at the lens surface. Thus, the antennas are easy to machine and comparatively inexpensive. Lenses have very large bandwidth and excellent wide-angle scanning properties. Furthermore, there is no aperture blockage by the feed system, and lens antennas can be designed to have low sidelobes (less than −35 dB) and a high front-to-back ratio. In the case of zoned lenses, however, the sidelobe level becomes frequency sensitive. In this chapter, millimeter wave characteristics are firstly summarized. Next discussed are two important applications: substrate lenses for imaging and submillimeter wave lenses for communications. Finally, millimeter wave lens antennas are analyzed in the last section.

4.1 MILLIMETER WAVE CHARACTERISTICS

This section presents benefits of millimeter wave technology (from 30 to 300 GHz) and its characteristics [2]. Utilization of this frequency band for the design of data transmission and sensing systems has a number of advantages:

1. The very large bandwidth resolves the spectrum crowding problem and permits communication at very high data rates.
2. The short wavelength allows the design of antennas of high directivity but reasonable size so that high-resolution radar and radiometric systems and very compact guidance systems become feasible.
3. Millimeter waves can travel through fog, snow, and dust much more readily than infrared or optical waves.
4. Finally, millimeter wave transmitters and receivers lend themselves to integrated and, eventually, monolithic design approaches, resulting in RF heads (front ends) that are rugged, compact, and inexpensive.

4.1.1 Millimeter Wave Loss Factors

Propagation effects have a strong influence on the design and performance of millimeter wave systems and, for this reason, are briefly reviewed here. As a general rule, millimeter wave transmission requires unobstructed line-of-sight (LOS) paths, but propagation into shadow zones is possible by edge diffraction and scatter, though at a reduced signal level. Factors that affect millimeter wave propagation are discussed below.

4.1.1.1 Free-Space Propagation. As with all propagating electromagnetic (EM) waves, for millimeter waves in free space, the power falls off as the square of range. For a doubling of range, power reaching a receiver antenna is reduced by a factor of four. This effect is due to the spherical spreading of the radio waves as they propagate. The frequency and distance dependence of the loss between two isotropic antennas can be expressed in absolute numbers by the following equation (in decibels):

$$L_{\text{free space}} = 20 \log_{10}\left(4\pi \frac{R}{\lambda}\right) (\text{dB}), \tag{4.1}$$

where $L_{\text{free space}}$ is the free-space loss, R is the distance between transmit and receive antennas, and λ is the operating wavelength. This equation describes LOS wave propagation in free space. This equation shows that the free-space loss increases when frequency or range increases. Also, millimeter wave free-space loss can be quite high even for short distances. It suggests that the millimeter wave spectrum is best used for short-distance communications links. The *Friis equation* gives a more complete accounting for all the factors from the transmitter to the receiver (as a ratio, linear units) [3]:

$$P_{\text{RX}} = P_{\text{TX}} G_{\text{RX}} G_{\text{TX}} \frac{\lambda^2}{(4\pi R)^2 L}, \tag{4.2}$$

where

G_{TX} is transmitter antenna gain;

G_{RX} is receiver antenna gain;

R is the LOS distance between transmit and receive antennas;

λ is wavelength (in the same units as R); and

L = system loss factor (≥ 1).

Most millimeter wave systems require antennas of high directivity/gain, and these antennas have been predominantly investigated. It is one of the advantages of the millimeter wave region that such antennas have very reasonable dimensions. Certain applications in millimeter wave guidance and radar also require fan-shaped radiation characteristics.

At millimeter wave frequencies, quasi-optical principles can be applied to the design of lens antennas as this type of antenna works in a similar manner for millimeter waves. In optical terminology, the index of refraction n is used in place of the dielectric constant ε_r:

$$n = \sqrt{\varepsilon_r}. \tag{4.3}$$

4.1.1.2 Atmospheric Losses.
Transmission losses occur when millimeter waves traveling through the atmosphere are absorbed by molecules of oxygen, water vapor, and other gaseous atmospheric constituents. These losses are greater at certain frequencies, coinciding with the mechanical resonant frequencies of the gas molecules.

Absorption by rain and atmospheric gases is the dominant effect, and it is evident that the choice of the operating frequency of a millimeter wave system will depend strongly on the desired transmission range. The H_2O and O_2 resonance have been studied extensively for purposes of predicting millimeter wave propagation characteristics. Figure 4.1 shows an expanded plot of the atmospheric absorption versus frequency at altitudes of 4 km and sea level, for a water content of 1 and 7.5 g/m³,

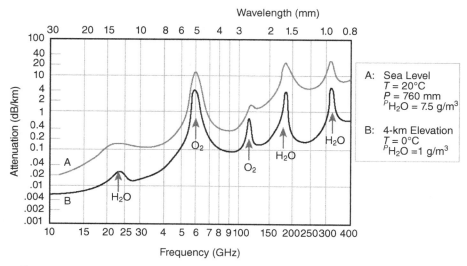

Figure 4.1. Average atmospheric absorption of millimeter waves (© 2005 IEEE [4]).

respectively (the former value represents relatively dry air, while the latter value represents 75% humidity for a temperature of 10°C). Large transmission distances can be obtained in the low-attenuation windows at 35, 94, 140, 220, and 340 GHz. On the other hand, operation near one of the steep absorption lines, for example, the O_2 line at 60 GHz, will allow one to control the transmission distance by adjusting the frequency. In this way, "overshoot" can be minimized and hence the probability of detection and interference by an unfriendly observer. Since the millimeter range, moreover, permits the design of antennas of high directivity and very small beamwidth, a system of optimum transmission security can be realized.

Another loss—proportional to the frequency squared—comes from the *Friis path loss* Equation (4.2). That "loss," however, is attributable to another factor. If omnidirectional antennas, such as half-wavelength dipoles, are used, then as the frequency rises, the effective area of the antennas decreases as frequency squared. If, on the other hand, the area of antennas is kept constant, then there is no increase in path loss because the electrical area increases as the wavelength decreases (squared).

For instance, a 60-GHz antenna, which has an effective area of 1 sq in., will have a gain of approximately 25 dBi, but this gain comes at the expense of being highly directional. This would mean that, for millimeter wave radios to be used at their full potential, they would need a solution for precise aiming.

4.1.1.3 Dielectric Losses.
Loss due to dielectric loss tangent (tan δ) is very important at millimeter wave frequencies. For many materials, it tends to be proportional to frequency, so the higher the frequency, the more likely it will be to dominate overall material loss. Table 4.1 lists loss tangent of materials suitable for the upper microwave spectrum for dielectric antennas and microstrip antennas. In the millimeter

TABLE 4.1. Millimeter Wave Materials

	Relative Dielectric Constant (ε_r)	Typical loss tangent (tan δ)
Teflon	2.08	0.0006
Polytetrafluoroethylene	2.17	0.0009
(PTFE) glass	2.33	0.0015
Polyethylene	2.26	0.0006
Duroid 5880	2.20	0.0006
Duroid 5870	2.33	0.0005
Quartz Teflon	2.47	0.0006
Polystyrene	2.54	0.0012
Fused quartz	3.78	0.000,25
Boron nitride	4.40	0.0003
Sapphire	9.0	0.0001
Alumina	9.8	0.0001
Silicon	11.7	0.002
Gallium arsenide	13.2	0.001
Magnesium titanate	16.1	0.0002

wave region (>30 GHz), the permittivity values should be approximately the same, but the loss factors may be higher.

4.1.1.4 Wave Impedance. The wave impedance of the material can be obtained from the index of refraction [5]:

$$Z = \sqrt{\frac{\mu}{\varepsilon}} = \frac{377}{n}, \tag{4.4}$$

where the impedance of free space is taken as 377 ohm.

When waves encounter an impedance discontinuity, they are partially transmitted and partially reflected in a similar manner to transmission line impedance mismatches. Also, when the radiation is incident at an angle other than perpendicular to the surface, refraction or bending of light EM radiation occurs in accordance with Snell's law. The "rays" are bent toward the surface normal when entering a medium with a higher dielectric constant and toward the surface normal when going from a higher to a lower dielectric constant material. (This is generally known as geometric optics.) This approach is fairly accurate for structures that are large compared to the wavelength of the radiation.

4.1.2 Ray-Tracing Propagation

This is best illustrated by example. For a silicon lens ($\varepsilon_r = 11.7$) without a matching layer, a typical reflection loss of 1.5 dB is reported in Reference 6, which implies that 30% of the power is reflected at the lens/air interface. In fact, these reflected rays are

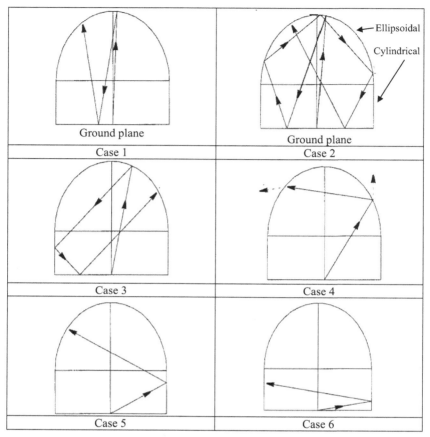

Figure 4.2. Second-order ray tracing of reflected waves inside the lens (©2001 IEEE [7]).

not lost; they eventually come out of the lens after multiple reflections inside the lens, reducing the directivity and contributing to the final radiation pattern.

Figure 4.2 shows second-order ray tracing of reflected waves inside the designed lens. In this work, second-order analysis is defined so that a ray is traced after it hits the lens surface for the first time and until it reemerges for a second time. For this analysis, the entire lens surface, including the ellipsoidal lens surface and cylindrical extension layer, can be divided into six different regions, each region corresponding to a different multiple reflection path.

Table 4.2 lists the corresponding source elevation angle for each region and its trace points on the lens/air interface before the ray exits the lens. For example, case 1 represents rays with source elevation angles between 0° and 2.65° and having all trace points for the first hit lying on the ellipsoidal surface. The transmitted waves are accounted for in the first-order ray tracing, as reported in References 6 and 8, and the reflected waves continue traveling inside the lens. The trace points for the second hit

TABLE 4.2. Ray Tracing of Reflection Waves for Different Regions [7]

Case	Source Elevation	First Hit	Second Hit	Third Hit	Fourth Hit
1	0°–2.65°	Ellipsoidal	Ground plane	Ellipsoidal	(Out)
2	2.65°–6.15°	Ellipsoidal	Ground Plane	Ellipsoidal	Ellipsoidal
3	6.15°–7.5°	Ellipsoidal	Ground plane/	Cylindrical/	Ellipsoidal (out)
	7.5°–12.5°		cylindrical	ground plane	
4	12.5°–51.1°	Ellipsoidal	Ellipsoidal	(Out)	–
5	51.1°–75.0°	Cylindrical	Ellipsoidal	(Out)	–
6	75.0°–90°	Cylindrical	Ellipsoidal	(Out)	–

lie on the ground plane where the rays get totally reflected. Finally, the rays hit the lens boundary for the third time on the ellipsoidal surface and they exit out of the lens. Other cases are similar to case 1 except case 2, for which the rays suffer multiple total internal reflections (5–10 times) before finding their way out of the lens.

From the point of view of ray tracing, the effect of the lens can be understood in terms of the culmination of the wave fronts emanating from the feeding patch antenna.

If we compare the front-to-back (F/B) ratio of the lens antennas, this is the ratio of the maximum directivity of the antenna to its directivity in the opposite direction. When the radiation pattern is plotted on a relative *decibel* scale, the front-to-back ratio is the difference in decibel between the level of the maximum radiation in the forward direction and the level of radiation at 180°.

In this framework, the relationship between the F/B ratio of the substrate lens in terms of the lens antenna directivity and the corresponding properties of the feed patch can be derived as follows. According to the definition, the F/B ratio of the patch and lens antennas, $(F/B)_{patch}$ and $(F/B)_{lens}$, respectively, can be written as

$$(F/B)_{patch} = U_{patch}^{forward} / U_{patch}^{backward} \text{ (in the dielectric half-space)} \tag{4.5}$$

$$(F/B)_{lens} = U_{lens}^{forward} / U_{lens}^{backward} \text{ (in air)}, \tag{4.6}$$

where U represents the maximum forward or backward radiation intensity of the patch or lens antenna. Furthermore, superscripts represent forward and backward radiation, whereas subscripts represent the patch and lens, respectively. The ratio between the F/B ratios and directivities for the patch and lens antennas can be calculated as

$$\begin{aligned}
\frac{(F/B)_{patch}}{D_{patch}} &= \frac{U_{patch}^{forward} / U_{patch}^{backward}}{4\pi U_{patch}^{forward} / P_{in}} = \frac{P_{in}}{4\pi U_{patch}^{backward}} \\
\frac{(F/B)_{lens}}{D_{lens}} &= \frac{U_{lens}^{forward} / U_{lens}^{backward}}{4\pi U_{lens}^{forward} / P_{in}} = \frac{P_{in}}{4\pi U_{lens}^{backward}},
\end{aligned} \tag{4.7}$$

where D_{patch} and D_{lens} are the directivities of the patch and lens antenna, respectively, and P_{in} is the common input power to the patch and lens. We now assume that the back radiation of the patch antenna is not significantly affected when replacing the infinite

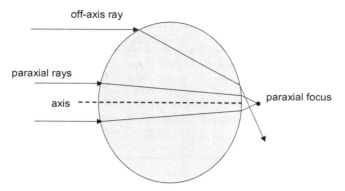

Figure 4.3. Spherical lens geometry.

dielectric half-space above the patch by a finite ellipsoidal lens with the same dielectric constant. This is a reasonable approximation since the lens is electrically large (four to five free-space wavelengths) and the ground plane provides good isolation between the front and back radiation. From Equation (4.7), and with $U_{\text{patch}}^{\text{backward}} \approx U_{\text{lens}}^{\text{backward}}$, we obtain

$$\frac{(F/B)_{\text{patch}}}{D_{\text{patch}}} = \frac{(F/B)_{\text{lens}}}{D_{\text{lens}}}. \tag{4.8}$$

This equation can be rewritten in decibels as

$$\left(\frac{F}{B}\right)_{\text{lens}} \approx \left(\frac{F}{B}\right)_{\text{patch}} + (D_{\text{lens}} - D_{\text{patch}}) \text{ in decibel.} \tag{4.9}$$

A theory of *spherical lenses* (Fig. 4.3) based on ray tracing has been confirmed by experiments [1]. The theory includes homogeneous lenses, the so-called constant-K lenses, as a special case. These latter lenses [9] should be made from materials with dielectric constants roughly in the range $2 \leq \varepsilon_r \leq 4$.

For $\varepsilon_r > 4.0$, the focal point would be located in the interior of the lens, which is undesirable, and for $\varepsilon_r < 2$, problems associated with large focal lengths may arise. The distance R' of the focal point from the center of a lens of radius R is approximately given by

$$\frac{R'}{R} = \frac{\sqrt{\varepsilon_r}}{2(\sqrt{\varepsilon_r} - 1)}, \tag{4.10}$$

which defines the position of the feed horn. Particularly for lenses with permittivity values close to the upper limit of 4, spherical aberrations are small and such lenses can be designed for diameters of up to 15 λ_0 and a gain of up to 30 dB while maintaining good pattern quality, including a sidelobe level below −20 dB. Spherical lenses are explored in much more detail in Chapter 6.

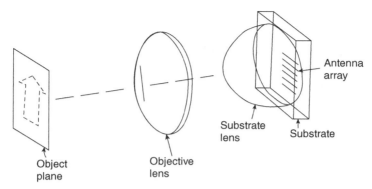

Figure 4.4. Operating principle of imaging array (© 1992 IEEE [10]).

4.2 MILLIMETER WAVE SUBSTRATE LENS FOR IMAGING

Substrate lens-coupled antennas are often applied to imaging arrays. In Figure 4.4, an image of an object is focused through an objective lens and a substrate lens onto an array of antennas and detectors. The signal detected at each antenna is plotted to form the image. The antennas are spaced at an interval that satisfies the Nyquist sampling criterion (typically two antennas per dielectric wavelength), so that the signals can be interpolated to recover the original image.

The concept of substrate lens has been commonly used because of its simple construction and robustness. The main advantage of the substrate lens is to minimize coupling effects by trapping surface waves in the substrate. In an imaging array, the objective lens projects the object plane onto the substrate surface (image plane), where the receiving array is located. Plotting the output of the antenna detectors produces the received image. In addition to serving as base plane for the array, the planar substrate has the beneficial effect of directing the radiating patterns of the array elements into the direction of the incoming waves. But surface waves supported by the planar substrate would counteract this effect by reducing the effective receiving cross section of the array elements and by increasing mutual coupling.

Absorption and reflection losses of the substrate lens are minimized by choosing a low-loss dielectric (the lens is made from the same material as the substrate) and by using a quarter-wavelength antireflection coating.

It has been noticed that an elementary antenna on a dielectric substrate can radiate most of its power into the substrate rather than into the air [11]. (Note that, in this context, elementary antennas comprise the family of electrically short dipoles, short slots, and small loops [magnetic dipoles], where the current can be accurately approximated as a constant or a linear distribution.) The ratio of the power in the air to that in the substrate is $(1/\varepsilon_r)^{1.5}$, where ε_r is the dielectric constant. Quartz is an insulator with a dielectric constant of 4, and silicon and gallium arsenide are semiconductors with higher dielectric constants between 12 and 13. This means that we would expect only $(1/4)^{1.5}$ or 13% of the power to propagate into space above a quartz substrate, and as little as $(1/13)^{1.5}$ or 2% for the silicon or gallium arsenide cases.

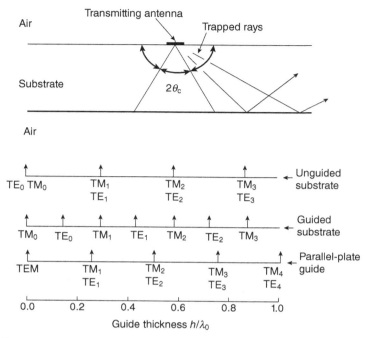

Figure 4.5. A transmitting antenna on a dielectric substrate and the associated substrate modes. All the energy radiated above the critical angle is trapped in the substrate and coupled to substrate modes. The critical angle shown is $\theta_c = 30°$, appropriate for fused quartz ($\varepsilon_r = 4$) (©1985 IEEE [11]).

Elementary antennas on planar dielectric substrates suffer from power loss to substrate modes (Fig. 4.5). In a transmitting antenna, the rays that are transmitted at angles larger than the critical angle are completely reflected and trapped as substrate modes. This power can be a significant fraction of the total power. By reciprocity, a receiving antenna suffers the same loss.

Detailed analyses of the surface-wave *losses* have been carried out by Pozar [12] and by Alexopoulos and coworkers [13–16]. From their results, it was observed that elementary antennas couple power to successively higher-order substrate modes as the thickness of the substrate increases, and in some cases, more than 90% of the radiated power is trapped in the dielectric [17].

The substrate mode issue can be minimized if the dielectric thickness is close to infinite; that is, the antenna is placed on a semi-infinite dielectric substrate. It was observed that dipole and slot antennas on infinite dielectrics can have a wideband (30–40%) input impedance (but not necessarily a pattern bandwidth) when operated in this region [10].

The "infinite" dielectric can be approximated using a lens of the same dielectric constant attached to the antenna substrate (in reality, because the lens is quite thick, it acts almost as well as an infinite dielectric layer above the transmission line substrate).

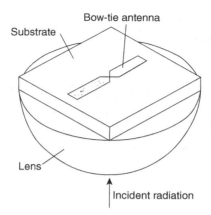

Figure 4.6. Substrate lens-coupled antenna (©1985 IEEE [11]).

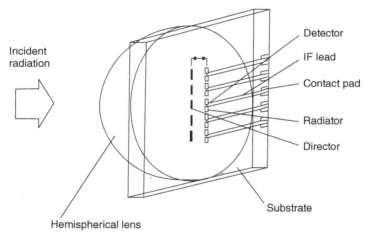

Figure 4.7. Yagi-Uda antenna imaging array. IF, intermediate frequency. (© 1992 IEEE [20].)

The lens is put on the back side of the substrate and focusing the incident radiation through the lens (Fig. 4.6). This takes advantage of the fact that the antenna responds primarily to radiation from the substrate side. The substrate modes are eliminated because the transmitted rays are now incident nearly normally on the lens surface. The disadvantages of the substrate lens are reflection loss and dielectric loss. The reflection loss is particularly important for a silicon lens because of its high dielectric constant. However, it is possible to eliminate this loss almost entirely with a polystyrene cap that acts as a matching layer [18].

A bow-tie antenna (Fig. 4.6) or a Yagi-Uda imaging array with high packing density was developed to work with a lens in References 19 and 20. In Figure 4.7, the concept was the use of a dipole antenna and a parasitic director element for improving

Figure 4.8. An elliptical lens superimposed on an extended hemispherical lens for different permittivity.

the radiation pattern of each of the elementary dipole antennas. The Yagi-Uda antenna was used in a 50-GHz plasma diagnostic experiment using a Teflon hyper-hemispherical lens. The beam can also be steered in one direction if the parasitic director is displaced relative to the active dipole. This has the potential of improving the coupling efficiency of the elements at the edge of the imaging array. Other researchers developed a substrate-backed microstrip antenna and a coplanar-waveguide slot antenna, and these have been measured at 52 and 11 GHz, respectively [20, 21]. The antennas show good radiation patterns and are compatible with microstrip and coplanar wave-guide feed networks. However, ohmic losses will limit their use to frequencies below 100 GHz [22].

The substrate lens can be a hemispherical, hyper-hemispherical, elliptical, or an extended hemispherical type (see Fig. 4.8). The substrate lens requires a matching layer to reduce reflection losses at the air–dielectric interface.

Kasilingam and Rutledge [23] reported the focusing properties of small substrate lenses and observed that the minimum radius for acceptable operation is 0.5 λ_0 for quartz and 1 λ_0 for silicon lenses. The hyper-hemispherical substrate lens was first used by Rutledge and Muha [24] at millimeter wave frequencies to eliminate the power loss to substrate modes and to increase the gain of planar antennas. It is an aplanatic lens; that is, if the optical system is designed such that all the rays are being focused to a point, the hyper-hemispherical lens can be added to the system and all the rays will still focus to a point (Fig. 4.9). The f number of a hyper-hemispherical substrate lens objective lens optical system f_s is related to the f number of the objective lens alone (f_o) by the equation

$$f_s = f_0 / n_1, \tag{4.11}$$

where n_1 is the index of refraction of the substrate lens.

The f number is defined using the subtended angle of the objective lens at the focal point by $f_s = 1 / \sin \theta_1$ and $f_o = l / \sin \theta_o$ [25]. The angle θ_1 is always larger than θ_0 and is given by

$$\frac{\sin \theta_1}{\sin \theta_0} = n_1, \tag{4.12}$$

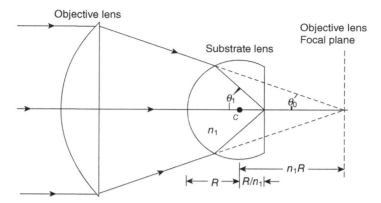

Figure 4.9. A hyper-hemispherical lens combined with an objective lens to form an imaging system (©1985 IEEE [11]).

which is the geometric relationship used to construct the hyper-hemispherical substrate. In antenna terms, the hyper-hemispherical substrate bends the rays radiated by the integrated antenna toward the broadside direction, thereby sharpening the pattern and effectively increasing the gain of the integrated antenna by n_1^2. This is equivalent to Equation (4.12) once it is realized that the sharpening of the pattern occurs in both the E- and H-planes.

We know that an elliptical lens focuses a paraxial wave (plane wave) to a point. The focal point is located at the second focus of the ellipse. The shape of the elliptical lens depends on the index of refraction of the lens used, and it is easy to derive the formula [26]. The elliptical lens has infinite magnification since a spherically diverging beam from the focal point is transformed into a plane wave. In antenna terms, this means any antenna placed at the focus of the elliptical lens will result in a far-field pattern with a main beam that is diffraction limited by the aperture of the elliptical lens. The difference between these antennas is then in the sidelobe and cross-polarization levels. Since the patterns are diffraction limited by the lens and therefore are very narrow, any increase in the sidelobe level can have a *detrimental* effect on the overall efficiency of the system. The elliptical lens is compatible with large-f-number imaging systems owing to its narrow diffraction-limited patterns, and should be placed near the minimum waist plane, where no phase errors are present in the Gaussian beam. This is in contrast to the hyper-hemispherical lens, which should be placed in a *converging* beam (i.e., with an appropriate phase error) for maximum coupling to an optical system.

In a hemispherical lens with a ground plane, it should be realized that the components of the electric field arise directly from the feed and lens and those that are reflected from the ground plane. The latter term contributes essentially to the main lobe of a spherical lens, while the former term constitutes a relatively small component. A continuous radial variation is difficult to achieve in practice, and so lenses of this type are usually constructed from a series of concentric shells: The topic is discussed in more detail in Chapters 6 and 7.

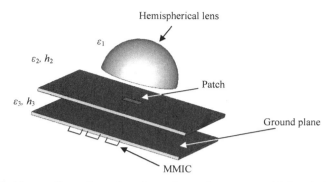

Figure 4.10. The configuration of a microstrip patch antenna with hemispherical lens.

4.3 MILLIMETER WAVE AND SUBMILLIMETER WAVE LENS

A hemispherical lens has better mechanical stability than spherical lens when the former is attached to a flat substrate. If we want to make two-dimensional arrays using the monolithic microwave integrated circuits (MMICs) technique, it is possible to have a lens-coupled patch antenna configuration as can be seen in Figure 4.10. This is compact and has advantages when including additional integrated circuits. This is particularly important in fabricating two-dimensional arrays [20, 27].

Such an array consists of two individual microstrip substrates separated by a common metal ground plane. The antennas are printed on the first substrate (ε_{r2}) covered with the low-loss dielectric lens (ε_{r1}). Each antenna is fed with a coupling slot [28, 29], from the MMIC constructed on the lower substrate (ε_3) .The lower substrate offers an efficient space for fabricating additional integrated circuits such as matching circuits, mixers, amplifiers, and interconnections. The antennas are ideally isolated from these circuits by the ground plane [30].

The patch is separated from MMICs by two substrates, and its size is subject to the relative permittivity for upper and lower substrates. If we define

$$\varepsilon_{12} = \frac{\varepsilon_2}{\varepsilon_1}. \tag{4.13}$$

As the ratio of the dielectric constant of the first substrate ε_2 to the dielectric constant of the lens ε_1, both the patch length a and the patch width b are

$$a = b = \frac{\lambda_{\text{eff}}}{2}, \tag{4.14}$$

where the effective wavelength in the first substrate λ_{eff} is defined by

$$\lambda_{\text{eff}} = \frac{\lambda_{\text{o}}}{\sqrt{\varepsilon_{\text{eff}}}} \tag{4.15}$$

and the effective dielectric constant of the first substrate ε_{eff} is given by [31]

$$\varepsilon_{\text{eff}} = \frac{\varepsilon_{12}+1}{2} + \frac{\varepsilon_{12}-1}{2}\left(1+\frac{10h_2}{b}\right)^{-1/2}, \tag{4.16}$$

where h_2 is the thickness of the first substrate and b is the patch width [31]. If ε_{r1} equals ε_{r2}, then the ratio ε_{12} becomes one, and then an ideal radiation pattern that is almost symmetrical for both the E- and H-planes can be realized. This pattern has neither any sidelobes nor radiation at the horizontal directions, which offers low cross talk and high coupling efficiency to the incident beam. When ε_{12} does not equal one, an undesirable substrate mode is generated in the first substrate, which affects the radiation and the impedance characteristics of the adjacent antennas in the array [32].

The next consideration is related to antenna mounting. Different methods are known for mounting a lens to a planar antenna or substrate. The conventional way for mounting the lens is to use additional mechanical holders [33], but these holders are heavy, expensive, and introduce additional reflections, which would affect the radiation pattern. Another possible solution is to add a foam sandwich layer between the lens and the patch itself. Since the foam has a dielectric constant close to that of air, it should not influence the performance of the antenna system. However, the multiple glue layers, which were needed to be applied, influenced the performance of the lens at millimeter wave frequencies. To overcome these problems, a novel "eggcup" type of lens can be constructed, which has a small size and is light weight. It is also easy to manufacture and therefore has a low cost [34].

The cross-sectional view of the lens is shown in Figure 4.11. It consists of a quasi lens, a waveguide, and a cavity. The lens can be designed using geometric optics methods. The cavity and waveguide also works as a lens supporter. To minimize its effect on the lens and the patch, both cavity and waveguide should have thin dielectric walls. All these parts should be made from the same dielectric material to keep good impedance matching.

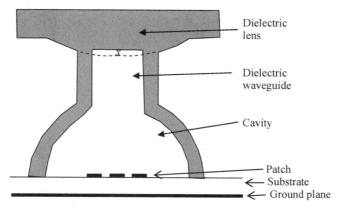

Figure 4.11. Cross-sectional view of "eggcup" lens on substrate. The dielectric structure consists of a parabolic-type lens, and the waveguide and a cavity act as the lens supporter (©2006 IEEE [34]).

The cavity is designed to contain the resonant energy, while the waveguide is designed to transform and filter the required mode to pass through (Fig. 4.11). A further increase of the cylinder diameter leads to multiple reflections within the cylinder and the performance is degraded. For easier manufacturing, we can deviate from the calculated lens contour within the cylindrical holder by employing a flat surface with negligible degradation in the performance (x point in Fig. 4.11).

The achieved half-power beamwidths for this arrangement are $20°$ in both the E-plane and the H-plane. The first sidelobes are below 15 dB. Measured and simulated gain of the complete antenna is around 15 dB over the frequency range from 57 to 63 GHz [33]. One should note that dispersion losses and tolerances of the dielectric constant of the materials may ultimately affect antenna performance.

4.3.1 Extended Hemispherical Lens

An extended hemispherical lens is like a hemielliptical lens and it performs the usual lens collimation function: to focus a plane wave to a point. Its principle is based on refraction at spherical surfaces. In physical optics, only the tangential electric and magnetic fields at the lens interface between the dielectric and free space are calculated. The Schelkunoff equivalence principle is then applied to substitute equivalent magnetic and electric currents for the surface magnetic and electric fields, respectively, and the radiation patterns are then computed from these equivalent currents. This is also known in optics as Babinet's principle.

A ray from an axial point S intersects the spherical surface at height h in Figure 4.12. After refraction, the ray converges and intersects the axis at a point F. The optical path length between S point and F point can be expressed as

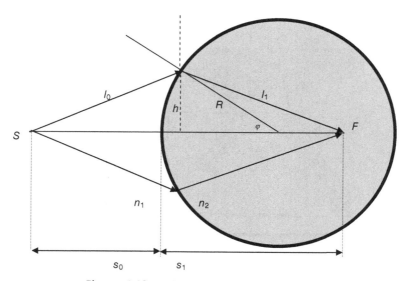

Figure 4.12. Refraction at a spherical surface.

$$n_1 l_0 + n_2 l_1$$
$$= n_1 \sqrt{R^2 + (s_0 + R)^2 - 2R(s_0 + R)\cos\phi} + n_2 \sqrt{R^2 + (s_1 - R)^2 - 2R(s_1 - R)\cos\phi},$$

$$(4.17)$$

where n_1 is the index of the air and n_2 is the index of the medium.

Using Fermat's principle and paraxial approximation, we can express refraction at spherical surfaces as follows

$$\frac{n_1}{s_0} + \frac{n_2}{s_1} = \frac{n_2 - n_1}{R}. \qquad (4.18)$$

If point S is located at a position that $s_0 \gg s_1$ (e.g., plane wave), the above equation can be simplified to

$$\frac{n_2}{s_1} \approx \frac{n_2 - n_1}{R} \text{ or } \frac{R}{s_1} + \frac{n_1}{n_2} \approx 1. \qquad (4.19)$$

When R and n_1 is fixed, we can find that the lens with the higher index n_2 (or higher dielectric constant) will have a smaller converging length s_1 (Fig. 4.12). The higher permittivity, the smaller the size the antenna can be, and these can be implemented using planar wafers. Extended hemispherical lenses can be synthesized with an ellipse, and it was shown in Reference 35 that the synthesized ellipse presented results in less than a 6% decrease in the Gaussian coupling efficiency at 500 GHz for a 6.8-mm silicon or quartz lens from a true elliptical lens.

When parallel rays entering a lens do not come to focus at a point, we say that the lens has an aberration. As can be seen in Figure 4.13, if light enters too large a region of a spherical surface, the focal points are spread out at the back. This is called spherical aberration. One solution for spherical aberration is to make sure that the diameter of any spherical lens is small in comparison to the radius of curvature of the lens surface.

For ease of fabrication of an extended hemispherical lens, the dimensions are chosen to approximate the desirable focusing properties of an elliptical lens with a feed located at one of its foci:

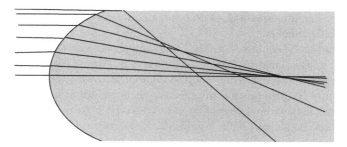

Figure 4.13. Spherical aberration.

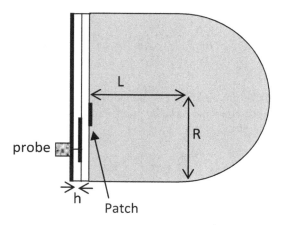

Figure 4.14. Cross-sectional view of proximity-coupled patch lens antenna (© 2001 IEEE [37]).

$$b = a\sqrt{\varepsilon_r / (\varepsilon_r - 1)}$$
$$d = b\sqrt{\varepsilon_r} \qquad\qquad (4.20)$$
$$L = d + b - R$$

where R is the lens radius at the maximum waist, b is the major semiaxis of elliptical curvature, d is the cylindrical extension length for an elliptical lens, and L is the total combined cylindrical extension for the extended hemispherical approximation.

Figure 4.14 shows a cross-sectional view of the proximity-coupled microstrip patch-lens configuration. Here, a microstrip line parasitically excites a rectangular microstrip patch. Above the patch is a dielectric lens terminating a cylindrical cross section of length L and radius R ($R = a$) (an extended lens). The lens is fabricated as an ellipsoidal lens to maximize the directivity [35]. As mentioned previously, to ensure no power is lost to surface waves, the grounded substrate for the feed line, the layer on which the microstrip patch is etched, and the material for the lens must have the same permittivity. As a further option, we can consider a proximity-coupled patch antenna on an extended hemispherical dielectric lens for millimeter wave applications (Fig. 4.14). The configuration has several advantages over the conventional microstrip antenna lens arrangements.

First, no surface-wave losses will be found associated with the feed network if the same dielectric constant materials are used for the multilayered patch configuration and the lens (as opposed to the aperture-coupled configurations in References 36 and 37). Using a proximity-coupled patch configuration yields greater bandwidths than a direct contact fed patch lens without degrading the front-to-back ratio of the antenna, unlike for aperture-coupled patches [36, 37], or printed slot versions [38]. Noncontact feeding techniques, such as proximity coupling, also tend to have lower cross-polarization levels than direct contact excitation methods [39].

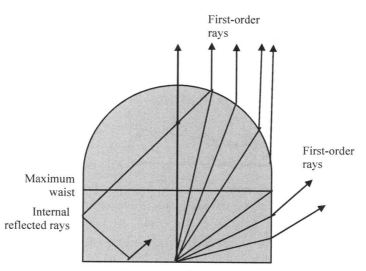

Figure 4.15. Dielectric lens modeling: 2-D ray tracing for an elliptical lens, with second-order internally reflected rays (© 2006 IEEE [43]).

As a further point, low-cost, low-dielectric constant materials such as polyethylene (PE) can be used without degradation of the front-to-back ratio, unlike a slot configuration [6, 37, 38, 40–42]. It is found that the aperture-coupled patch approach in Figure 4.16 leads to the least spurious radiation from the feed lines or electronics integrated close to the patch antenna. In the case of the proximity-coupled patch antenna, this immunity to parasitic radiation is compromised [35].

The size of a lens should ensure that the lens surface is located in the far field of the printed feed's radiation pattern for both the "first-order" rays, which have a single point of intercept with the lens surface as shown in Figure 4.15, and the internally reflected rays, which are called "second-order." A radius of 12.5-mm lens with a permittivity of 12.0 can be used as a starting point for a 60-GHz scale model [37].

As can be seen in Figure 4.15, the lens's collimating property is only effective over its rounded surface above the plane of its maximum waist. Geometric optics analysis reveals that feed radiation intercepting the lens surface below the maximum waist, at the surface of a cylindrical extension, is not collimated but rather propagates laterally in undesired directions. For this reason, the most efficient feed architectures for use with lens antennas should be designed to minimize radiation in lateral directions along the ground plane. Such lateral radiation can also be found for other feed architectures such as the conventional dual-slot feed [35] and the twin arc-slot design [40].

The behavior modeling of this antenna can be simplified by assuming that the radius and length of the extended lens are significantly greater than the dimensions of the patch antenna. Following this design rule allows the microstrip antenna to be represented as if it were mounted in an infinite half-space of dielectric constant ε_r, which greatly simplifies the analysis required. This assumption has been used in several publications [37, 38] to model an aperture-coupled patch lens antenna.

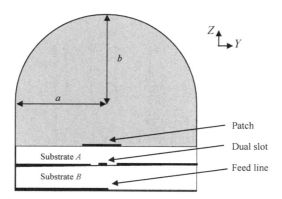

Figure 4.16. Layout for the aperture-feed ellipsoidal substrate lens.

In order to determine the radiation performance of the lens antenna, the radiation pattern emanated by the patch into the dielectric lens can be calculated from the currents on the patch [44]. This radiation will illuminate the spherical surface of the lens. The far field can be computed based on the equivalent surface electric current density and the equivalent surface magnetic current density on the spherical surface of the lens [37].

When the cost of lens is a concern, a medium-cost polymer such as PE should be suitable. Cross-linked polystyrene such as Rexolite (a trade name) is even better as it exhibits superior machinability, albeit at a higher cost than PE. Both materials exhibit low loss at millimeter wave frequencies [45]. Applying the design method in Reference 37 and using $\varepsilon_r \approx 2.35$, the length of the lens is 64 mm and the radius 50 mm for an operation centered at 60 GHz. Substrate A in Figure 4.16 is preferred to have the same permittivity as the lens to minimize the surface-wave losses. The design methodology for a proximity-coupled patch in this environment is similar to that when mounted in free space [46, 47],. Thus, for a given set of dielectric materials, the resonant frequency is governed by the length of the patch and the impedance at resonance is controlled by and the offset of the terminated feed line from the center of the patch. Since a low dielectric constant material was used, it was deemed unnecessary to coat the lens with an antireflection layer [48].

A general schematic diagram of the single-beam substrate lens antenna is shown in Figure 4.16. A lens of this type has been made out of Rexolite.

For generating diffraction-limited patterns, an ellipsoidal lens ($x^2/a^2 + y^2/a^2 + z^2/b^2 = 1$) may be chosen, with $a = b\sqrt{(\varepsilon_r - 1)/\varepsilon_r}$, where a and b are the minor and major axes of the ellipsoidal lens, respectively. However, the extension length beyond the major axis is cylindrical instead of elliptical to facilitate the machining process. According to geometric optics, the length of the cylindrical extension layer should be equal to $b/\sqrt{\varepsilon_r}$ in order to generate parallel rays through the lens when the feed antenna is located on the axis at the far focal point of the ellipsoidal lens. The radiating element used to feed the lens is realized by an aperture-coupled circular polarized (CP) patch antenna (Fig. 4.16).

The feed line of the antenna is built on a high-permittivity substrate B, and the patch antenna is printed on a low-permittivity substrate A, which is close to the permittivity of the Rexolite lens. These choices for the substrate are made to increase the bandwidth, as well as to reduce the parasitic radiation losses due to the feed network.

The main advantage of this aperture-coupled patch antenna is that the feeding network and the radiating element are well separated by a ground plane and, thus, the patterns are immune to parasitic radiation [49]. Also, the ground plane yields an increased front-to-back (F/B) ratio, which is important since low-permittivity materials are used. Another advantage is that the single line feed structure is well suited for IC applications.

If circular polarization is needed, it can be made by means of a circular-polarized patch or a cross-shaped slot in the ground plane, which excites two orthogonal modes in a nearly square patch [50]. In particular, the cross aperture-coupled structure was reported to yield a significant improvement to CP bandwidth. Research shows that the circular polarization properties of the structure are not very sensitive to manufacturing tolerances.

For a silicon lens ($\varepsilon_r = 11.7$) without a matching layer, a typical reflection loss of 1.5 dB is reported in Reference 6, which implies that 30% of the power is reflected at the lens/air interface. In fact, these reflected rays are not lost; they eventually come out of the lens after multiple reflections inside the lens, reducing the directivity and contributing to the final radiation pattern.

4.3.2 Off-Axis Extended Hemispherical Lens

The dielectric lens also provides mechanical rigidity and thermal stability and has been used extensively in millimeter and submillimeter wave receivers [51–56]. In Section 4.3.1, it has been mentioned that if the dielectric lens has the same dielectric constant as the planar antenna wafer, then substrate modes can be eliminated [41, 57]. In addition, antennas placed on dielectric lenses tend to radiate most of their power into the lens side, making the pattern unidirectional on high-dielectric constant lenses. The ratio of powers between the dielectric and air is approximately $\varepsilon_r^{3/2}$ for elementary slot and dipole-type antennas [57] where ε_r is the relative dielectric constant of the lens.

Research works [51] have shown that the directivity of the substrate lens can be controlled by changing the extension length L, defined in Figure 4.17. In particular, as the extension length increases from the hyper-hemispherical length R/n (where R is the radius and n is the index of refraction of the lens), the directivity increases until it reaches a maximum diffraction-limited value.

While the directivity increases at higher extension lengths, the pattern-to-pattern coupling value to a fundamental Gaussian beam (Gaussicity) decreases [35]. A Gaussian beam propagating along the z-axis as in Figure 4.18 produces a propagating field as

$$E = \frac{E_0}{z - jz_0} \exp(-jkr^2 / 2(z - jz_0)),$$ (4.21)

where $r = (x^2 + y^2)^{1/2}$ and E_0 is a constant.

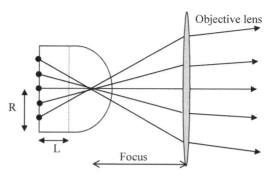

Figure 4.17. A simplified linear imaging array on an extended hemispherical dielectric lens-coupled to an objective lens (© 1997 IEEE [6]).

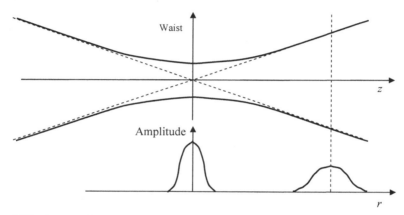

Figure 4.18. Gaussian beam amplitude variations versus distance to the axis follow a Gaussian law, the width of which increases with z, while its amplitude decreases with z.

If the double-slot antenna used in Figure 4.16 launches a nearly perfect fundamental Gaussian beam into the dielectric lens, the Gaussicity can also be thought of as a measure of the aberrations introduced by the lens.

When extension lengths are up to the hyper-hemispherical position, the Gaussicity is nearly 100% since the hyper-hemispherical lens is aplanatic, implying the absence of spherical aberrations, and satisfies the sine condition, which guarantees the absence of a circular coma [58]. As the extension length L increases past R/n, the Gaussicity continuously decreases, which implies the introduction of more and more aberrations. Research results show that for an "intermediate position" between the hyper-hemispherical and diffraction-limited extension lengths (e.g., $L/R = 0.32$–0.35 for silicon lens), the Gaussicity decreases by a small amount ($<10\%$), while the directivity is close to the diffraction-limited value [35]. The choice of an intermediate position extension length has resulted in state-of-the-art receivers at 90 and 250 GHz [8, 53, 54].

Figure 4.17 shows the off-axis performance of extended hemispherical dielectric lenses. A ray optics/field-integration formulation can be used to solve for the radiation patterns and Gaussian coupling efficiencies. Briefly, the radiation of the feed antenna is ray traced to find the fields immediately exterior to the lens surface. For a given ray, the fields are decomposed into TE/TM components at the lens/air interface, and the appropriate transmission formulas are used for each mode. The equivalent electric and magnetic currents are found directly from the fields, and a standard diffraction integral results in the far-field lens patterns [59].

In most applications, the dielectric lens will be coupled with a quasi-optical system, and Figure 4.17 shows the dielectric lens coupled to an objective lens. If the Gaussian beams emanating from the dielectric lens are well characterized, then one can easily trace these beams through a quasi-optical system [35], or for greater accuracy, the patterns emanating from the dielectric lens could be used with EM ray-tracing techniques to find the fields across the aperture of the objective lens. Then, a Fourier transform will yield the far-field patterns from the objective lens/dielectric lens system.

Any antenna that illuminates the lens surface with a nearly symmetrical, constant phase beam will produce similar results. The black circle in Figure 4.17 represents a radiation element such as the dipole in Figure 4.19 or a dual slot. The array radiation can be calculated by assuming a sinusoidal magnetic current distribution on the dipole/slot and using an array factor in the E-plane direction [60]. The dimensions of the double-slot antenna are a length of 0.28 λ_{air} and spacing of 0.16 λ_{air} for a silicon lens with $\varepsilon_r = 11.7$. The dimensions can be scaled to other dielectric material using the square root of the dielectric constant.

The wavelength of the sinusoidal magnetic current distribution in the slot is approximately the geometric mean wavelength given by $\lambda_m = \lambda_0 / \sqrt{\varepsilon_m}$, where $\varepsilon_m = (1+\varepsilon_r)/2$ [61]. If the double-slot antennas produce a radiation pattern with 98% Gaussicity, the dielectric lens should also have a similar radiation pattern unless aberrations are introduced by the lens. Note that the patterns radiated to the air side are broader and contain 9.0% of the total radiated power for a silicon lens. The theoretical technique for analyzing the lens-radiation patterns is an expanded version of the EM ray-tracing technique presented in Reference 35.

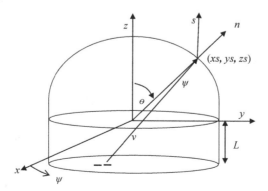

Figure 4.19. The dipole-feed lens geometry used for the off-axis theoretical computations.

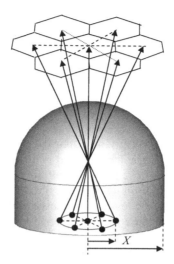

Figure 4.20. Multiple-beam launching through substrate lens antenna (© 2001 IEEE [37]).

The lens antenna can be used to launch multiple beams by printing multiple antenna elements under the base of the lens [6, 37]. As can be seen in Figure 4.20, an array at the back of the lens is used to provide efficient coverage. The scan angle depends on the off-axis displacement X/a, where X is the off-axis distance in Figure 4.20 and a is the minor axis of the designed ellipsoidal lens. For wireless communications, one of the most important features for multiple-beam antennas is scan coverage. As demonstrated in Reference 6, the off-axis total internal reflection loss is the limiting factor in the design of larger multiple-beam arrays on substrate lenses. For the present circular polarization design, another possible limitation is off-axis depolarization.

The peak directivity drops quickly as off-axis displacement increases. In order to launch beams with equal radiation power density and to reduce reflection losses, the effect of the extension length L has been numerically investigated [37] and the optimum position has been found to lie around $L \approx a/\sqrt{\varepsilon_r}$. This seems to correspond to the "intermediate" position previously observed for extended hyper-hemispherical lenses [6, 35].

4.3.3 Submillimeter Wave Lens Antennas for Communications

For submillimeter wave front ends, corrugated horn antennas are used very often, but they are getting more expensive to manufacture and the effect of misalignments becomes more severe when the frequency increases. By applying modern technology in nanostructuring and micromachining, one can make reliable production and alignment of planar antenna structures on dielectric lenses with accuracy sufficient for the submillimeter wave range.

Lens antenna with photomixing (or optical heterodyne down-conversion) is one of the conventional configurations. The photomixing process is commonly used to

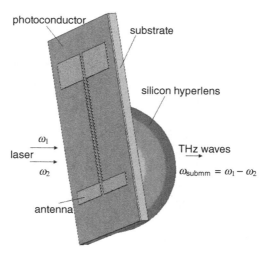

photoconductor

substrate

silicon hyperlens

ω_1

laser

ω_2

THz waves

$\omega_{submm} = \omega_1 - \omega_2$

antenna

Figure 4.21. A schematic view of two-beam photomixing with a photomixer. Two-beam photomixing with a photomixer couples to a silicon hemispherical lens. The incident angular frequencies are ω_1 and ω_2, respectively. The output submillimeter wave frequency ω_{submm} is the difference between ω_1 and ω_2 (© 2011 IEEE [62]).

generate submillimeter wave continuous waves from two continuous-wave laser beams. We define that these lasers have angular frequencies ω_1 and ω_2, respectively. Both beams with the same polarization are mixed and then focused onto an ultrafast semiconductor material. Due to the photonic absorption and the short charge carrier lifetime in the material, we obtain the modulation of the conductivity at the expected submillimeter wave frequency $\omega_{submm} = \omega_1 - \omega_2$. Figure 4.21 schematically shows the two-beam photomixing with a photomixer coupled to a hemispherical substrate lens. Figure 4.22 shows a schematic diagram of the photoconductive emitter and an identical detector.

To detect submillimeter wave pulse signals, the waves first are focused on a detector or a photoconductive antenna. The double refraction of a silicon lens (Fig. 4.21) or an electro-optic crystal can be read using a probe pulse, usually split up from the laser used in submillimeter wave generation. The probe beam is then measured with a quarter-wavelength plate. Finally, the time delay between the probe pulse and the submillimeter wave signal can be measured to obtain the electric field in the time domain.

For submillimeter wave pulse emission and detection, the femtosecond optical pulses from a mode-locked Ti:sapphire laser are focused by an objective lens on the biased gap of the photoconductive antenna (see Fig. 4.23), mounted on the flat side of a silicon hyper-hemispherical lens, to pump and generate photocarriers. This excitation gives rise to a transient current and emits a submillimeter wave pulse. The silicon lens reduces the loss caused by the reflection and refraction of radiation at the surface/air interface. The emitted submillimeter wave pulse is collimated and focused by a pair of off-axis paraboloidal mirrors onto a sample first, then the diverging submillimeter wave

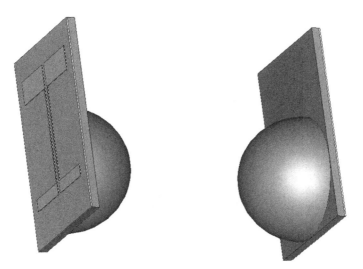

<u>Figure 4.22.</u> A schematic diagram of the photoconductive emitter and identical detector along with a typical free-space submillimeter wave system incorporating lenses. The same arrangement can be applied to both pulsed and continuous wave-based submillimeter wave communications systems (© 2011 IEEE [62]).

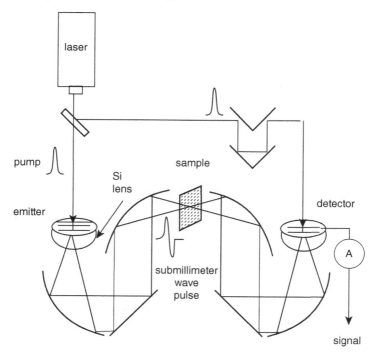

<u>Figure 4.23.</u> Optical setups for submillimeter wave pulse emission and detection, with Si hyper-hemispherical lenses in an emitter and a detector.

beam from the sample point is collimated and then focused by a pair of off-axis parabo-loidal mirrors onto a photoconductive antenna detector mounted on the back of the silicon hyper-hemispherical lens with the same diameter. That is why lens antennas play key roles in submillimeter wave communications.

4.4 ANALYSIS OF MILLIMETER WAVE SPHERICAL LENS

Spherically symmetric lenses of a simple structure are capable of providing very good electrical performance. Practical requirements on spherically symmetric lenses include a very uniform permittivity throughout the lens and correctness in shape to avoid beam tilting and pattern deformations. Dielectric losses in constant-K lenses can be estimated from the formula [63]

$$L_{\text{diel}} = 36 \frac{r}{\lambda_0} [\varepsilon_r^{3/2} - (\varepsilon_r - 1)^{3/2}] \tan \delta, \tag{4.22}$$

where $tan \, \delta$ is the loss factor of the lens material and L is in decibel.

For a quartz lens with a diameter of 15 λ_0 and $\tan \delta = 0.00025$, the loss would be 0.18 dB. A similar loss would occur in a 10 λ_0 Teflon lens with $\tan \delta = 0.0006$. In the case of a two-layer lens consisting of a core and shell, the step in refractive index at the outer surface is, in general, large enough to require an antireflection layer. A single matching layer can be effective up to large angles of incidence [1]. The step between core and shell is usually sufficiently small so that reflection is negligible, and no match-ing layer is needed at the inner lens surface.

Shaping techniques are developed for millimeter wave lens antennas [64, 65]. Speaking in general terms, the design procedure imposes the power conservation law to control the aperture taper. Snell's law is used to define the first surface of the lens as the radiation source is located at the feed point. Next, we can determine the second surface by considering the path length condition. The path length condition is specified by the desired phase distribution across the aperture. The radiation pattern of the lens is calculated in the usual way with scatter effects at the zoning edges neglected.

However, these lens shaping methods differ from the well-established design tech-niques for optical lenses [66]. Optical correction techniques are concerned primarily with spherical lenses and the compensation of their phase errors (aberrations), while the objective of millimeter-band shaping techniques is the realization of a desired phase and amplitude distribution across the lens aperture.

For multibeam antennas, a method, termed *coma-correction zoning*, is used to minimize distortions of off-axis beams caused by cubic phase errors [64]. The basic idea is that coma aberrations of a thin lens are significantly reduced provided the inner surface of the lens, that is, the surface facing the feed horn, is spherical on the average. This condition can be satisfied by zoning. Lenses that combine contour shaping with coma-correction zoning are of particular interest for use in millimeter wave satellite communications systems. This application requires spaceborne multibeam antennas capable of radiating a large number of closely spaced, highly directive beams of high

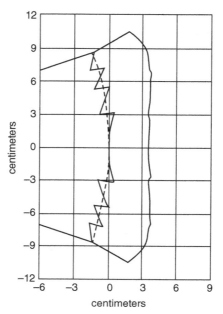

Figure 4.24. Cross-section profile of coma-reduced, shaped lens (focal length = 24.5 cm) for multibeam operation at 44 GHz (© 1983 IEEE [66]).

pattern quality, which implies minimum pattern degradation for off-axis beams and low sidelobes (to allow frequency reuse). In addition, a high crossover level of 4 dB is required for adjacent beams in order to ensure continuous ground coverage.

Figure 4.24 shows the cross-section profile of a shaped, coma-reduced lens designed by Lee and Carlise [66]. The lens has a focal length of 24.5 cm, an aperture diameter of 20.7 cm = 30 λ_0, and a center thickness of 3.5 cm = 5 λ_0 at the design frequency of 44 GHz. The feed pattern is assumed to be the standard E-plane pattern of a square horn; that is,

$$g(\theta) = (1 + \cos\theta)\frac{\sin[(\pi d / \lambda_0)\sin\theta]}{(\pi d / \lambda_0)\sin\theta}, \tag{4.23}$$

where θ is the angle counted from the horn axis. The horn size d is chosen to provide an edge taper of 30 dB at $\theta = 20°$, the aperture half-angle of the lens as seen from the feed location. The lens is designed to produce an aperture distribution $E(r) = [1 - (r/1.05)^2]^3$ with a uniform phase. Made from Rexolite with $\varepsilon_r = 2.54$, the lens has a weight of 0.5 kg.

An experimental model of the lens of Figure 4.24 has yielded a gain of 33 dB, a beamwidth of 3.3°, and a sidelobe level of −30 dB for individual beams [64]. These beams could be scanned up to ±12° without significant coma degradation. For multibeam operation, the lens can be fed by a cluster of small single-beam feed horns generating closely spaced beams with the desired high crossover level of 4 dB. The aperture efficiency of the antenna is 48%. Total losses amount to 2–3 dB.

Satellite communications systems will require larger antennas of a much higher directivity in many cases, but this experimental version demonstrates the excellent performance that can be obtained through the use of lens shaping methods.

Beam steering lens antennas can be designed by employing variable permittivity media in the form of electric field-controlled liquid artificial dielectrics [67]. Two approaches for the realization of liquid artificial dielectrics media have recently been pursued, that is, forming solutions of high-molecular-weight rigid macromolecules and suspending highly asymmetric micrometer-sized metallic particles in a low-loss base liquid. Using the second approach, variable phase shifters consisting of discrete liquid artificial dielectric-filled cells in metal waveguides have been demonstrated at 35 and 94 GHz [67]. The experiments suggest that with careful attention to cell tolerances, a 360° phase shift could be achieved with a control voltage in the order of 100–200 V and losses as low as 1.8–2.4 dB. Applied to antennas, liquid artificial dielectrics media would permit the design of steerable lenses whose local electrical thickness is controlled with the help of a system of electrodes distributed throughout the lens volume [67]. In this way, a uniform phase taper across the aperture could be affected for beam deflection, or any desired phase distribution could be achieved for adaptive beam correction. The feasibility and practicality of liquid dielectric lenses as millimeter wave antennas remain to be established. Nonetheless, the number of useful liquid artificial dielectrics media is probably very large, and only very few have been studied to date [67]. (Some of those explored so far were both flammable and toxic!)

4.5 WAVEGUIDE-FED MILLIMETER WAVE INTEGRATED LENS

A lens that is mechanically coupled directly to a planar feed network is attractive at millimeter wave frequencies because of compactness and structural integrity—a separate supporting structure between the feed and lens is not needed. These are the majority of feed strategies considered in the above sections. However, a similar approach may be taken with a waveguide feed. In Reference 68, a WR15 waveguide feed was used to address the 60-GHz band, the cross section of the waveguide being just 1.9 × 3.8 mm. Into this was inserted a single dielectric component, which comprised a corrugated lens along with an integrated tapered waveguide insert, the latter providing the impedance match required between what would otherwise be an abrupt discontinuity between the air-filled waveguide and the dielectric. The dielectric used was alumina. The fabrication process was stereolithography rather than computer numerical control (CNC) machining. In stereolithography, components are built up layer by layer via laser polymerization. It is considered a "rapid prototyping" process and one that is fully automated by a dedicated machine, and as such, is very useful for the fabrication of millimeter wave components in small production runs [69].

The lens antenna of Nguyen et al. [68] very closely resembled that illustrated in Figure 4.25, where the metal components are lightly shaded so that the alumina part can be viewed. In measurements that were carried out on the fabricated antenna, the

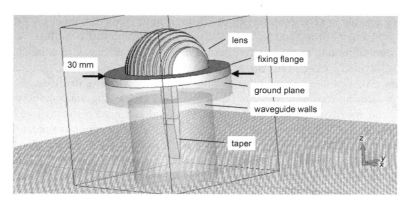

Figure 4.25. Integrated waveguide-fed lens for 60 GHz.

reflection coefficient was better than −10 dB, and radiation efficiency was between 60% and 86% from 55 to 65 GHz. Midband gain was 19 dBi. Further literature on integrated lens antennas is found in Reference 70.

REFERENCES

[1] T. L. ApRhys, "The Design of Radially Symmetric Lenses," IEEE Trans. Antennas Propag. Vol. AP-18, July 1970, pp. 497–506.

[2] K. Huang and D. J. Edwards, Millimeter Wave Antennas for Gigabit Wireless Communications, John Wiley & Sons, Chichester, 2008.

[3] H. T. Friis, "A Note on a Simple Transmission Formula," Proc. IRE Vol. 34, 1946, pp. 254–256.

[4] M. Marcus and B. Pattan, "Millimeter Wave Propagation; Spectrum Management Implications," IEEE Microwave Magazine Vol. 6, No. 2, 2005, pp. 54–62.

[5] G. Chartier, Introduction to Optics, Springer Science+Business Media, Berlin, 2005.

[6] D. F. Filipovic, G. Gauthier, S. Raman, and G. M. Rebeiz, "Off-Axis Properties of Silicon and Quartz Dielectric Lens Antennas," IEEE Trans. Antennas Propag. Vol. 45, 1997, pp. 760–766.

[7] X. Wu, G. V. Eleftheriades, and T. E. van Deventer-Perkins, "Design and Characterization of Single- and Multiple-Beam mm-Wave Circularly Polarized Substrate Lens Antennas for Wireless Communications," IEEE Trans. Microwave Theory Tech. Vol. 49, No. 3, 2001, pp. 431–442.

[8] D. Filipovic, G. V. Eleftheriades, and G. M. Rebeiz, "Off-Axis Imaging Properties of Substrate Lens Antennas," Fifth International Symposium on Space Terahertz Technology, Ann Arbor, MI, pp. 778–787, 1994.

[9] W. Free, F. Cain, et al., "High-Power Constant-Index Lens Antennas," IEEE Transactions on Antennas and Propagation, Vol. 22, 1974, pp. 582–584.

[10] G. M. Rebeiz, "Millimeter-Wave and Terahertz Integrated Circuit Antennas," IEEE Proc. Vol. 80, 1992, pp. 1748–1770.

[11] D. B. Rutledge, "Substrate-Lens Coupled Antennas for Millimeter and Submillimeter Waves," IEEE Antennas Propag. Soc. Newslett. Vol. 27, 1985, pp. 4–8.

[12] D. M. Pozar, "Considerations for Millimeter Wave Printed Antennas," IEEE Trans. Antennas Propag. Vol. 31, 1983, pp. 740–747.

[13] N. G. Alexopoulos, P. B. Katehi, and D. B. Rutledge, "Substrate Optimization for Integrated Circuit Antennas," IEEE Trans. Microwave Theory Tech. Vol. 31, 1983, pp. 550–557.

[14] P. B. Katehi and N. G. Alexopoulos, "On the Effect of Substrate Thickness and Permittivity on Printed Circuit Dipole Properties," IEEE Trans. Antennas Propag. Vol. 31, 1983, pp. 34–38.

[15] D. R. Jackson and N. G. Alexopoulos, "Gain Enhancement Methods for Printed Circuit Antennas," IEEE Trans. Antennas Propag. Vol. 33, 1985, pp. 976–987.

[16] D. R. Jackson and N. G. Alexopoulos, "Microstrip Dipoles on Electrically Thick Substrates," Int. J. Infrared Millimeter Waves Vol. 7, 1987, pp. 1–26.

[17] R. L. Rogers and D. Neikirk, "Use of Broadside Twin Element Antennas to Increase Efficiency on Electrically Thick Dielectric Substrates," Int. J. Infrared Millimeter Waves Vol. 9, 1988, pp. 949–969.

[18] C. Zah and D. Rutledge, "A Polystyrene Cap for Matching a Silicon Lens at Millimeter Wavelengths," Int. J. Infrared, Millimeter Waves Terahertz Waves Vol. 6, No. 9, 1985, pp. 909–917.

[19] Y. Daiku, K. Mizuno, and S. Ono, "Dielectric Plate Antenna for Monolithic Schottky-diode Detectors," Infrared Phys. Vol. 18, 1978, pp. 679–682.

[20] K. Uehara, K. Miyashita, K. Natsume, K. Hatakeyama, and K. Mizuno, "Lens-Coupled Imaging Arrays for the Millimeter-and Submillimeter-wave Region," IEEE Trans. Microwave Theory Tech. Vol. 40, 1992, pp. 806–811.

[21] D. H. Evans and P. J. Gibson, "A Coplanar Waveguide Antenna for MMICs," Proceedings of the 18th European Microwave Conference, pp. 312–317, 1988.

[22] D. Neikirk, "Integrated Detector Arrays for High Resolution Far-Infrared Imaging," PhD dissertation, California Institute of Technology, Pasadena, 1983.

[23] D. Kasilingam and D. Rutledge, "Focusing Properties of Small Lenses," Int. J. Infrared Millimeter Waves Vol. 7, No. 10, 1986, pp. 1631–1647.

[24] D. B. Rutledge and M. Muha, "Imaging Antenna Arrays," IEEE Trans. Antennas Propag. Vol. 30, 1982, pp. 535–540.

[25] M. Bom and E. Wolf, Principles of Optics, Pergamon Press, New York, 1959.

[26] E. Hecht and A. Zajac, Optics. Reading, Addison-Wesley, Boston, 1980, Chapter 5.

[27] G. M. Rebeiz, D. Kasilingam, Y. Guo, P. A. Stimson, and D. B. Rutledge, "Monolithic Millimeter-Wave Two-Dimensional Horn Imaging Arrays," IEEE Trans. Antennas Propag. Vol. 38, 1990, pp. 1473–1482.

[28] D. M. Pozar, "Five Novel Feeding Techniques for Microstrip Antennas," Antennas and Propagation Society International Symposium Digest, pp. 920–923, 1987.

[29] M. I. Aksun, S. Chuang, and Y. T. Lo, "On Slot-Coupled Microstrip Antennas and Their Applications to CP Operation-Theory and Experiment," IEEE Trans. Antennas Propag. Vol. 38, 1990, pp. 1224–1230.

[30] J. R. James, P. S. Hall, and C. Wood, Microstrip Antenna Theory and Design, Peter Peregrinus Ltd., London, 1981, Chapter 4.

[31] M. V. Schneider, "Microstrip Lines for Microwave Integrated Circuits," Bell System Tech. J. Vol. 48, 1969, pp. 1421–1444.

[32] A. K. Bhattacharyya, "Characteristics of Space and Surface Waves in a Multilayered Structure," IEEE Trans. Antennas Propag. Vol. 38, 1990, pp. 1231–1238.

[33] U. Sangawa, K. Takahashi, T. Urabe, H. Ogura, and H. Yabuki, "A Ka-band High-Efficiency Dielectric Lens Antenna with a Silicon Micromachined Microstrip Patch Radiator," in IEEE MTT-S Int. Dig., Vol. 1, pp. 389–392, 2001.

[34] M. Al-Tikriti, S. Koch, M. Uno, and A. Compact, "Broadband Stacked Microstrip Array Antenna Using Eggcup-Type of Lens," IEEE Microwave Wireless Compon. Lett. Vol. 16, No. 4, 2006, pp. 230–232.

[35] D. F. Filipovic, S. S. Gearhart, and G. M. Rebeiz, "Double-Slot Antennas on Extended Hemispherical and Elliptical Dielectric Lenses," IEEE Trans. Microwave Theory Tech. Vol. 41, 1993, pp. 1738–1749.

[36] G. V. Eleftheriades, Y. Brand, J.-F. Zurcher, and J. R. Mosig, "ALPSS: A Millimeter-Wave Aperture-Coupled Patch Antenna on a Substrate Lens," Electron. Lett. Vol. 33, 1997, pp. 169–170.

[37] X. Wu, G. V. Eleftheriades, and E. Van Deventer, "Design and Characterization of Single and Multiple Beam mm-Wave Circularly Polarized Lens Antennas for Wireless Communications," IEEE Antennas and Propagation Society International Symposium, Orlando, FL, pp. 2408–2411, 1999.

[38] P. Otero, G. V. Eleftheriades, and J. R. Mosig, "Integrated Modified Rectangular Loop Slot Antenna on Substrate Lenses for Millimeter- and Submillimeter-Wave Frequencies Mixer Applications," IEEE Trans. Antennas Propag. Vol. 46, 1998, pp. 1489–1497.

[39] D. M. Pozar, "Microstrip Antennas," Proc. IEEE Vol. 80, January 1992, pp. 79–91.

[40] M. Qiu and G. V. Eleftheriades, "Highly Efficient Unidirectional Twin Arc-Slot Antennas on Electrically Thin Substrates," IEEE Trans. Antennas Propag. Vol. 52, No. 1, 2004, pp. 53–58.

[41] D. F. Filipovic and G. M. Rebeiz, "Double-Slot Antennas on Extended Hemispherical and Elliptical Quartz Dielectric Lenses," Int. J. Infrared Millimeter Waves Vol. 14, 1993, pp. 1905–1924.

[42] J. Zmuidzinas and H. G. LeDuc, "Quasi-optical Slot Antenna SIS Mixers," IEEE Trans. Microwave Theory Tech. Vol. 40, 1992, pp. 1797–1804.

[43] A. P. Pavacic, D. L. del Rio, J. R. Mosig, and G. V. Eleftheriades, "Three-Dimensional Ray-Tracing to Model Internal Reflections in Off-Axis Lens Antennas," IEEE Trans. Antennas Propag. Vol. 54, No. 2, 2006, pp. 604–612.

[44] D. M. Pozar, "Radiation and Scattering from a Microstrip Patch on an Uniaxial Substrate," IEEE Trans. Antennas Propag. Vol. AP-35, 1987, pp. 613–621.

[45] P. F. Goldsmith, Quasi-Optical System, IEEE Press, Piscataway, NJ, 1998, p. 82.

[46] D. M. Pozar and S. M. Voda, "A Rigorous Analysis of a Microstripline-Fed Patch Antenna," IEEE Trans. Antennas Propag. Vol. 35, 1987, pp. 1343–1349.

[47] D. M. Pozar and B. Kaufman, "Increasing the Bandwidth of a Microstrip Antenna by Proximity Coupling," Electron. Lett. Vol. 23, 1987, pp. 368–369.

[48] L. Mall, Student Member, IEEE, and R. B. Waterhouse, "Millimeter-Wave Proximity-Coupled Microstrip Antenna on an Extended Hemispherical Dielectric Lens," IEEE Trans. Antennas Propag. Vol. 49, No. 12, 2001, pp. 1769–1772.

[49] X. Wu, G. V. Eleftheriades, and E. van Deventer, "A mm-Wave Circularly Polarized Substrate Lens Antenna for Wireless Communications," Symposium on Antenna Technology and Applied Electromagnetics (ANTEM), Ottawa, Canada, pp. 595–598, 1998.

[50] T. Vlasits, E. Korolkiewicz, A. Sambell, and B. Robinson, "Performance of a Cross-Aperture Coupled Single Feed Circularly Polarized Patch Antenna," Electron. Lett. Vol. 32, No. 7, 1996, pp. 612–613.

[51] T. H. Buttgenbach, "An Improved Solution for Integrated Array Optics in Quasioptical Millimeter and Submillimeter Waves Receivers: The Hybrid Antenna," IEEE Trans. Microwave Theory Tech. Vol. 41, 1991, pp. 1750–1761.

[52] J. Zmuidzinas, "Quasioptical Slot Antenna SIS Mixers," IEEE Trans. Microwave Theory Tech. Vol. 40, 1991, pp. 1797–1804.

[53] G. P. Gauthier, W. Y. Ali-Ahmad, T. P. Budka, D. F. Filipovic, and G. M. Rebeiz, "A Uniplanar 90 GHz Schottky Diode Millimeter Wave Receiver," IEEE Trans. Microwave Theory Tech. Vol. 43, 1995, pp. 1669–1672.

[54] S. S. Gearhart and G. M. Rebeiz, "A Monolithic 250 GHz Schottky Diode Receiver," IEEE Trans. Microwave Theory Tech. Vol. 42, 1994, pp. 2504–2511.

[55] H. Z. Zirath, C.-Y. Chi, N. Rorsman, and G. M. Rebeiz, "A 40-GHz Integrated Quasi-Optical Slot HFET Mixer," IEEE Trans. Microwave Theory Tech. Vol. 42, 1994, pp. 2492–2497.

[56] A. Skalare, H. van de Stadt, T. de Graauw, R. A. Panhuyzen, and M. M. T. M. Dierichs, "Double-Dipole Antenna SIS Receivers at 100 and 400 GHz," in Proceedings of the 3rd International Conference on Space Terahertz Technology, Ann Arbor, MI, pp. 222–233, 1992.

[57] D. B. Rutledge, D. Neikirk, and D. Kasilingam, "Integrated Circuit Antennas" in Infrared and Millimeter-Waves, Vol. 10, K. J. Button, ed., Academic, New York, 1983, pp. 1–90.

[58] M. Born and E. Wolf, Principles of Optics, Permagon, New York, 1959, pp. 252–252.

[59] C. A. Balanis, Antenna Theory: Analysis and Design, Wiley, New York, 1982, Chapter 11.

[60] R. S. Elliott, Antenna Theory and Design, Prentice-Hall, Englewood Cliffs, NJ, 1981, Chapter 4.

[61] M. Kominami, D. M. Pozar, and D. H. Schaubert, "Dipole and Slot Elements and Arrays on Semi-infinite Substrates," IEEE Trans. Antennas Propag. Vol. AP-33, 1985, pp. 600–607.

[62] K. Huang and Z. Wang, "Terahertz Terabit Wireless Communication," IEEE Microwave Magazine, pp. 108–116, June 2011.

[63] Y. T. Lo and S. W. Lee, Antenna Handbook: Antenna Applications, Vol. 3, Springer, Berlin, 1993.

[64] P. J. Kahrilas, "HAPDAR—An Operational Phased Array Radar," Proceedings of the IEEE, pp. 1967–1975, Nov. 1968.

[65] W. Patton, "Limited Scan Arrays," Proceedings of the 1970 Phased Array Antenna Symposium, Artech House Inc., Dedham MA, pp. 332–343, 1970.

[66] J. Lee and R. Carlise, "A Coma-Corrected Multibeam Shaped Lens Antenna, Part II: Experiments," IEEE Trans. Antennas Propag. Vol. 31, 1983, p. 216.

[67] H. T. Buscher, "Electrically Controllable Liquid Artificial Dielectric Media," IEEE Trans. Microwave Theory Tech. Vol. MTI-27, May 1979, pp. 540–545.

[68] N. T. Nguyen, R. Sauleau, N. Delhote, D. Baillargeat, and L. L. Coq, "Design and Characterization of 60-GHz Integrated Lens Antennas Fabricated through Ceramic Stereolithography," IEEE Trans. Antennas Propag. Vol. 58, No. 8, August 2010, pp. 2757–2762.

[69] K. F. Brakora, J. Halloran, and K. Sarabandi, "Design of 3-D Monolithic MMW Antennas Using Ceramic Stereolithography," IEEE Trans. Antennas Propagat. Vol. 55, No. 3, March 2007, pp. 790–797.

[70] M. J. M. van der Vorst, "Integrated Lens Antennas for Submillimeter Wave Applications," PhD dissertation, Department of Electronic Engineering., Eindhoven University of Technology, Eindhoven, The Netherlands, ISBN 90-386-1590-6, 1999.

5

LENS ANTENNAS FOR COMMUNICATIONS FROM HIGH-ALTITUDE PLATFORMS

John Thornton

This chapter discusses how lens antennas can be useful for certain applications where very low sidelobes are sought, or where a shaped beam is sought, in particular for millimeter wave frequencies. The experimental lens antennas discussed in this chapter arose as a consequence of research into communications from high-altitude platforms. Accordingly, this topic will be introduced and discussed in at least a little depth since it underpins the rationale for the study into the variants of lens antennas that follows.

5.1 INTRODUCTION

The following sections of the chapter review several research programs, or studies at a systems level, which gave rise to the author's interest in lens antennas as suitable candidates for these systems. In particular, a recurring theme encountered during research into communications using high-altitude platforms was the properties of the payload antennas. Here, multiple spot beams were cited as a route to exploiting very high spectral efficiency and thus to unlocking the commercial potential of the aerial communications concept. At that time, around 2000, much interest (some of it in

Modern Lens Antennas for Communications Engineering, First Edition. John Thornton and Kao-Cheng Huang.

retrospect perhaps better described as hyperbole) surrounded an expectation in a continued, exponential growth of wireless communications markets. The then recent auction of spectrum for third-generation mobile had netted unexpected billions of dollars for national governments, and a catchphrase of the times, at least among the communications community, might have been "spectrum is money."

To this end, a new phase of interest in HAPs—not by then a new concept [1]—spurred European Union (EU) investment, via the EU Commission, in some modest research projects. Aside from the noncommunications engineering content, central to these studies were antenna properties, propagation modeling, effects of rain and interference, and international radio regulations. In particular, HAPs were seen as being able to better exploit large bandwidths available at millimeter wave frequencies around 48 GHz, or later, 28–30 GHz, in a way not possible for terrestrial networks. For these historical reasons, much of this chapter is concerned with millimeter wave systems and antennas. Variants of lens antennas emerged as promising candidates for both payload and ground user antennas, for scanning antennas, and eventually for satellite communications. Scanning lens antennas for satellite communications are covered in Chapter 7.

5.2   THE HIGH-ALTITUDE PLATFORM CONCEPT

The basic concept [1, 2] is one where a vehicle, or a "platform" of some kind, is deployed in the higher reaches of the atmosphere and used to support wireless communications services. The altitude most typically cited would be in the region 17–23 km, firmly classified as being within the stratosphere. Proponents of the concept would typically go on to say that HAPs can offer the best features of terrestrial and satellite communications while avoiding the worst features. (Of course, opponents might take the contrary view, but for present purposes, that argument will not be further developed.) To this end, Figure 5.1 illustrates how HAP's relative proximity to the ground increases power flux density, compared to that offered by satellite, and allows for high-spectrum reuse using spot beam antennas.

Figure 5.2 then shows a chief difference between terrestrial and HAP wireless communications networks. In the former, propagation effects tend to be dominated by the built environment (illustrated) or the local ground terrain. Of course, the radio frequency being considered will also strongly influence propagation loss, but the trend of interest here is that when higher frequencies are considered (above a few gigahertz, up to and including millimeter waves), the issue tends to reduce to whether or not a line of sight (LOS) exists. Thus, in the terrestrial case shown in Figure 5.2a, several LOS links are found in close proximity to the base station, but shadowed regions with nonline of sight (NLOS) can be quite extensive, especially with increasing distance from the base station. Considering the HAP case of Figure 5.2b, the greatly increased elevation angle vastly increases the regions experiencing LOS. Regions of NLOS still exist, but these reduce in area, and for suburban or rural areas, the coverage offered by the HAP may be very extensive indeed, extending to 100 km or so [3, 4]. The HAP is thus often cited as being like a very low satellite or a very tall mast.

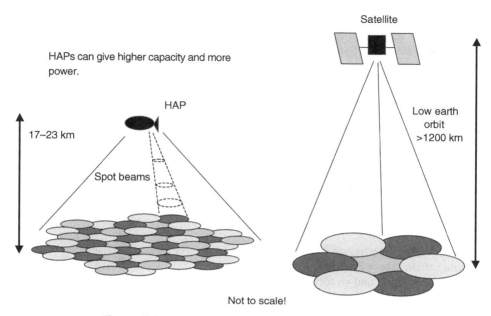

Figure 5.1. HAPs communications compared to satellite.

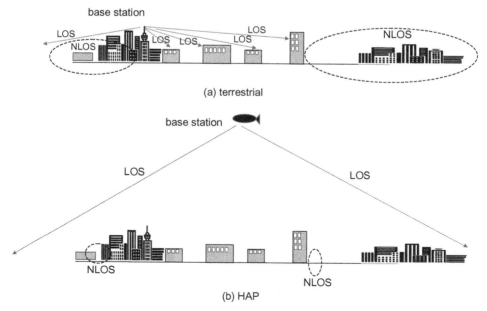

Figure 5.2. Propagation geometry in terrestrial and HAP communications.

At present, a detailed analysis of propagation effects is not being attempted nor any particular results put forward. Rather, the rationale behind the research programs such as the European *HeliNet* [5] and *CAPANINA* [6, 7] projects of the early-to-mid-2000s is introduced. Nevertheless, an excellent treatise on elevation angle-dependent shadowing is presented in Reference 8, where frequencies up to about 5 GHz are considered for standards such as IEEE 802.16x ("WiMAX"). A striking conclusion of Holis and Pechac [8], which appeared a little after the above projects had concluded, was that shadowing loss due to the built environment can be very severe indeed, both in terms of mean propagation loss and its statistical variance. This finding tends to cement the earlier supposition that establishing LOS to as many customers as possible is indeed key to exploiting spectrum and returns to the theme of how HAPs with multibeam antenna payloads could help to bring this about.

5.2.1 Spectrum Reuse Using HAPs

Thus far, the properties of HAP communications have been discussed in a rather general, qualitative way. Having begun with a stated interest in spectrum reuse using spot beam antennas and so forth, it becomes necessary at some point to apply detailed analytical methods so as to investigate system performance. Spectrum is reused, geographically, using cellular networks, and so a conventional cellular pattern should be one of our starting points. Figure 5.3 illustrates a layout of hexagonal cells where the channel reuse number is 4. Thus, four different patterns of shading are used to indicate cochannel cells (regions where the same channel is used). A channel in this context is a subdivision of the total spectrum asset, which can be divided by frequency (frequency division multiple access), time slot (time division multiple access) [9, 10], or by spreading code in a spread spectrum network (code division multiple access) [11]. Also, hybrid schemes that employ a combination of these techniques are encountered [10].

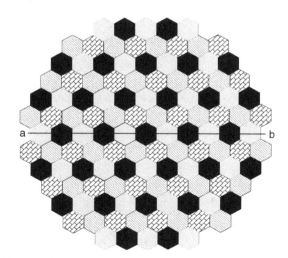

Figure 5.3. Hexagonal cells and four-channel reuse plan.

The next step should then be to ask how the radio frequency (RF) power would be distributed across this region of interest or service coverage area. A desired output will be the levels of cochannel interference, and so signal strength on the ground will first need to be derived. Now, whereas in terrestrial propagation it is terrain and buildings that exert a strong influence, and these are usually approached with statistical models, for HAPs (and having assumed LOS conditions), it is then the antenna radiation patterns that will dominate. This leads to the corollary question: What assumptions should be made about these radiation patterns? Being mindful of the HAP geometry in Figure 5.1, and the desired hexagonal cells of Figure 5.3, it is apparent that there tends not to be a single answer as to what should be the required antenna beamwidth. If a constant cell size is desired (as illustrated), then more distant cells will require illumination by payload antennas with a smaller beamwidth since the cell subtends a smaller angle as viewed from the HAP base station. Further, since these cells subtend *different* angles in azimuth and elevation, there would be some danger of becoming entangled in complexity before the first tentative steps toward analytical modeling have been taken. These problems were addressed in References 5 and 4, and where the following methodology was adopted:

- Assume a regular hexagonal pattern (as in Fig. 5.3) since this achieves tessellation and so tends to maximize coverage (avoids gaps in service).
- Calculate the required azimuth and elevation beamwidths for each cell.
- Assume a simple mathematical curve to represent the antenna main lobe.
- There will be an elevation plane main lobe pattern and a pattern for the orthogonal azimuth plane. The three-dimensional pattern can be derived from interpolation between these two curves.
- Assume a flat sidelobe floor.

These assumptions are not necessarily of high accuracy if we attempt to reconcile them with real antenna radiation patterns (e.g., sidelobe regions are usually anything but flat) but do at least lead to a tractable starting point. In particular, it will later be shown how sidelobe structure is of much less importance than the mean sidelobe power when considering systems of many cochannel beams; hence, the term "system level" was coined in the introduction to this chapter.

The "simple mathematical curve" alluded to above could, of course, take many forms.

A popular starting point might be of the form $\sin(\sin\theta)/(\sin\theta)$, or the "sinc" function.

In a fuller form, the power in the far field of a uniform rectangular aperture is given by

$$P(\theta) = \left(\frac{\sin\left(\pi \dfrac{D}{\lambda}\sin\theta\right)}{\pi \dfrac{D}{\lambda}\sin\theta} \right)^{2}, \tag{5.1}$$

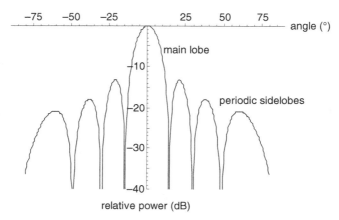

Figure 5.4. Radiation pattern of uniform aperture of length four wavelengths.

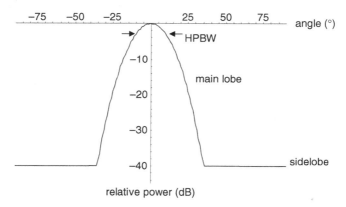

Figure 5.5. Simplistic radiation pattern with flat sidelobe region.

where D is the aperture dimension in the plane of the observation angle θ and λ is the wavelength. Such a pattern is shown in Figure 5.4.

Figure 5.4 exhibits the periodic sidelobes characteristic of the uniformly illuminated aperture. This could indeed be viewed as a more physically representative approach to that stated above, namely, the use of a flat sidelobe floor. However, the cost then is one where it becomes more difficult to unravel the effects of antenna beamwidth, cell angular spacing, and the way structured sidelobes from the many cochannel beams combine together to generate interference. While these effects are all worth studying in a later, more detailed analysis, it was argued in References 5 and 4 that the simpler radiation pattern model, that is, where the sidelobe region is flat (unvarying with angle), is a more pragmatic starting point.

Figure 5.5 shows a curve of the form

$$P(\theta) = (\cos\theta)^n; \tag{5.2}$$

that is, the main lobe power rolls of with angle according to Equation (5.2) until the level reaches a given sidelobe power relative to peak antenna gain: This sidelobe level is −40 dB in the example shown in Figure 5.5. In Equation (5.2), the term n is chosen to determine the rate of roll-off with angle. This approach is mathematically convenient because it is very easy to derive n for a given half-power beamwidth (HPBW) by

$$n = \log_{\frac{\mathrm{HPBW}}{2}}\left(\frac{1}{2}\right) \qquad (5.3)$$

since the power has rolled off by a factor of two at one-half of the full HPBW:

$$\left(\cos\frac{\mathrm{HPBW}}{2}\right)^n = \frac{1}{2}. \qquad (5.4)$$

In Figure 5.5, the HPBW is 20°, giving rise to $n = 45$. Thus, for any arbitrary HPBW, a curve can be derived using very few computational steps, an advantage when many different beamwidths may be required to optimally illuminate cells at different subtended angles as viewed from the HAP payload. These subtended angles are labeled θ_{sub} and ϕ_{sub} in Figure 5.6, respectively, the elevation and azimuth subtended angles as viewed from the aerial platform. Here, the platform height is h; the ground distance to the cell center is d_g; and θ_p is the pointing angle in elevation, along which the payload antenna illuminating the cell is aligned. Typical values would be $h = 20$ km and $d_g = 0$–30 km, although much greater ground distances could be studied, albeit leading to a regime where elevation angles become very low. The cell diameter is a further variable. Increasing the number of cells leads to smaller cells and higher spectral efficiency, which is an overreaching objective, but at the cost of increased antenna aperture area owing to the requirement to reduce beamwidths. A model in mind here is that, in

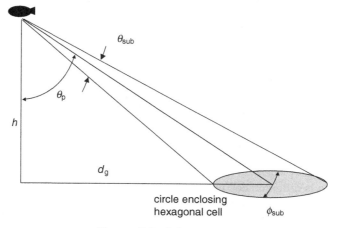

Figure 5.6. Cell geometry.

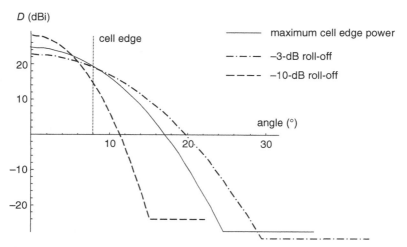

Figure 5.7. Effect of choice of cell edge antenna power.

the simplest case, a single cell is served by a single antenna. This has been called the "dedicated beam model." Later, multibeam antennas will be discussed.

A means of iteratively deriving the subtended angles, for both the cell centers and edges, is given in Reference 4. The algebra encountered is not particularly taxing, but the iterative method presented is useful when applied to computer code whose purpose may be to derive the power footprints for several hundreds of cells on the ground.

It is also important to choose a beamwidth (or more properly: two orthogonal beamwidths) that yields as uniform as possible a distribution of power and interference across each cell. Too narrow a beam will concentrate power in cell centers at the cost of excessive power roll-off toward cell edges, while too wide a beam gives rise to reduced power at cell centers and excessive cochannel interference due to beam overlap. The effect is shown in Figure 5.7, where three choices of radiation pattern are juxtaposed against an arbitrary cell edge angle of 8° (the cell subtends ±8° as viewed from the HAP).

Here, the example of a 10-dB power roll-off at the cell edge disadvantages users toward the edge of the cell, although we would expect cochannel interference to be potentially reduced, depending on the angular separation of cochannel cells, as a consequence of reducing off-axis radiation. A 3-dB roll-off increases cell edge power compared to the 10-dB case but is still not an optimal choice. By deriving a main lobe beamwidth that maximizes edge-of-cell power, a slight improvement on the simplistic 3-dB roll-off is manifested as increased power across the whole cell. Thornton et al. [4] present a method where an expression for cell edge directivity as a function of the roll-off factor n is differentiated to determine its maximum. However, in practice, the optimum HPBW so determined for many cell locations was found to be slightly less than the subtended cell angles, and that a fit of about 4.5 dB roll-off at the cell edge was usually a good working assumption.

5.2.2 Example Results: Cell Power and Interference

Having established some simple recipes for modeling a system of many different antenna spot beams, we can extract a good deal of output data relating to the characteristics of the cellular network. In particular, it is possible to derive power footprints on the ground and from these to then derive carrier-to-interference ratio, an important metric in any cellular network.

The logical flow is as follows:

- Determine ground coordinates of cochannel cells.
- Calculate subtended angles of cell edges.
- Calculate antenna beamwidths (the effect of varying the beamwidth can easily be explored).
- Iteratively, for each beam, derive power distribution on the ground as a function of radiation pattern and free-space distance. (If an asymmetric beam is used, it is necessary to use interpolation to model the three-dimensional pattern.)
- Derive carrier-to-interference ratio as a function of ground coordinates.

$$\mathrm{CIR}(x, y) = \frac{P\_\mathrm{wanted}(x, y)}{\sum P\_\mathrm{unwanted}(x, y)}. \tag{5.5}$$

Working with one of the four cochannel groups of cells from Figure 5.3 (of which there are 30 cells in total), and choosing one cell, on which the spot beam boresight is centered, three cases are next illustrated.

The beamwidths which fulfill the criterion of maximizing edge-of-cell power are in this example 7.6° in elevation and 11.6° in azimuth. The diameter of a circle enclosing a hexagonal cell is 6.3 km.

Figure 5.8 shows power contours for one spot beam, which illuminates a single cell where the azimuth and elevation beamwidths are both equal to the "best fit" to the cell's subtended elevation angle of 7.6°. This leads to an excessive power roll-off in azimuth and so disadvantages users at the cell edge.

In contrast, Figure 5.9 shows power contours where the azimuth and elevation beamwidths are again equal but chosen for the best fit to the cell's subtended azimuth angle of 11.6°. This leads to reduced power in the cell center and too slow a power roll-off in elevation, which tends to increased cochannel interference.

From Figures 5.8 and 5.9, it is evident that a symmetric beam for any cell other than one directly below the HAP will produce noncircular power contours on the ground. If the beam along one axis is fitted to the cell edge, the orthogonal beam is either too wide or too narrow. The power contours resemble ellipses, although strictly they are not. For cells near the center of the coverage area, the effect is slight and might be expected to have minimal effect on the quality of service and coverage, but for more distant cells, that is, those at lower elevation angles, the distortion can become problematic and there is more advantage in using asymmetric beams, as shown in Figure 5.10.

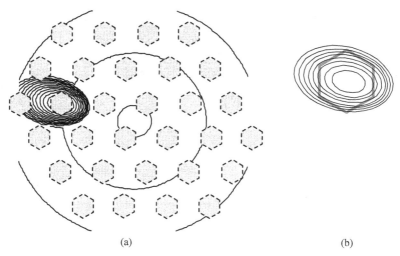

(a)   (b)

Figure 5.8. Spot beam power contours: symmetric beam with elevation fit. (a) Three-decibel contour spacing and (b) zoom in to cell with 1-dB contour spacing.

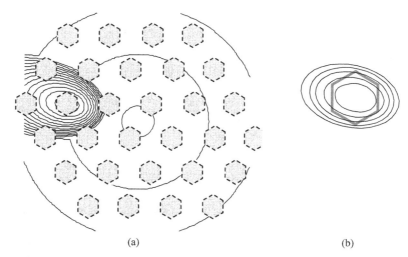

(a)   (b)

Figure 5.9. Spot beam power contours: symmetric beam with azimuth fit. (a) Three-decibel contour spacing and (b) zoom in to cell with 1-dB contour spacing.

Figure 5.10 shows power contours where the spot beam HPBWs are respectively 7.6° in elevation and 11.6° in azimuth, leading to near-circular cell footprints and the highest possible edge-of-cell power as explored in Figure 5.7.

Thus far, qualitative results have been presented, showing expected trends for cell power footprints resulting from various choices for antenna beamwidths. Now, let us assume that the power distribution on the ground has been calculated for each cochannel

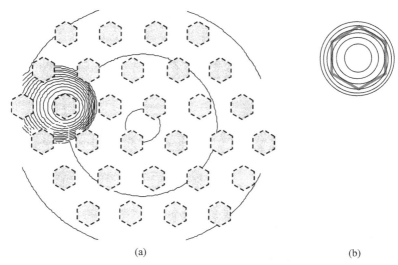

(a)   (b)

Figure 5.10. Spot beam power contours: optimized asymmetric beam. (a) Three-decibel contour spacing and (b) zoom in to cell with 1-dB contour spacing.

spot beam (or cell). This gives rise to a set of matrices for power as a function of Cartesian coordinates. In the example under consideration, there are 30 cells sharing the same channel, and so the entire cochannel network can be characterized by a set of 30 such matrices. It is then straightforward to derive the carrier-to-interference ratio (CIR) and, in a development of Equation (5.5), this can be defined as

$$\text{CIR}(x, y) = \frac{P_{\max}(x, y)}{\left(\sum_{i=1}^{N} P_i(x, y)\right) - P_{\max}(x, y)}. \tag{5.6}$$

In Equation (5.6), $P_{\max}(x, y)$ is the maximum power from the set of N antenna beams. Since only one spot beam is directed at the cell of interest, this also represents the wanted signal or "carrier." Then, the denominator in Equation (5.6) represents the summation of the power in all the other beams (all the beam powers added together minus the wanted power)—this is the cochannel interference.

Assuming that the user's link budget is adequate to support a given data rate, CIR is the next most important parameter. Indeed, cellular networks, conceived to reuse spectrum as much as possible, should be interference limited rather than power limited.

Figure 5.11a shows the CIR contours associated with the cell for which the power contours were shown in Figure 5.10. The CIR contours for the entire cochannel cell group would be a superposition of 30 such matrices, from which still further information may be extracted such as a statistical representation of CIR coverage. This is shown in Figure 5.11b, where we see that the simplistic circular beam recipe for spot beam beamwidths (as used in Fig. 5.8), while yielding a higher CIR in some small part of

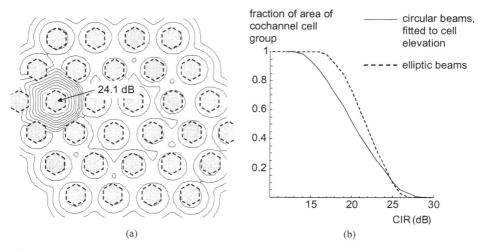

(a)   (b)

Figure 5.11. Cochannel carrier-to-interference ratio. (a) CIR contours for the cochannel cell of Figure 5.10. (b) Cumulative distribution function for the cochannel group: circular versus elliptic beams.

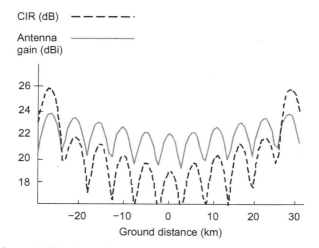

Figure 5.12. Linear distribution of CIR and power (gain) values.

the coverage area, is inferior to the elliptic beam recipe when considering the cell area as a whole. In particular, the elliptic beam recipe leads to a higher minimum CIR and is inferior to the circular beam recipe only for less than roughly 15% of the coverage area.

Cross-sectional distributions of power and CIR values through the center of the 60-km-diameter coverage area are shown in Figure 5.12: This cut is taken along the line ab in Figure 5.3. Here, the apparent antenna gain "as seen" on the ground is scaled for free-space loss, and it is apparent that this is a nearly constant ripple series of curves as antenna directivity rolls off on either side of each cell center. This near flatness is a consequence of increasing antenna directivity for more distant cells, effectively countering the additional free-space loss. However, as well as this, apparent antenna gain tends to increase slightly to the edges of the plotted curve because of the increasing ellipticity of the antenna beams, which is needed to accommodate the decrease in the subtended area of the more distant cells. A further subtlety could be pointed out: A 4.5-dB center-to-edge power roll-off is not apparent because the plotted curves are aligned along the flat sides of the hexagonal cells (Fig. 5.3), which fall inside the circular −4.5-dB contour.

Unlike the near-constant cell power distributions, CIR values are seen to vary much more widely. This is caused by several effects: the reduced number of neighbor cells at the region's outer edge; the increased free-space loss, which attenuates aggregate sidelobe power for these outer cells; and the scaling of the sidelobe floor, which is relative to peak antenna gain rather than as an absolute dBi quantity.

Many other relationships can be studied, such as the effect of raising or lowering the sidelobe floor, and a comparison of different channel reuse plans. To investigate the latter, it is necessary only to derive the Cartesian coordinates of the set of cells of interest. Further studies have shown how overlapping areas of coverage shared by different channels could aid handover (user movement between cells) and how the antenna payload might be steered on a gimbal to accommodate platform movement [12].

While it would be beyond the scope of the chapter to go into an increasing level of detail into studying these effects, relationships, and trade-offs, much can be found in literature. A conclusion which may be drawn at this stage is that much insight can be obtained from quite simple inputs and assumptions, and key to these is those pertinent to the payload antenna radiation patterns.

5.3 ADVANTAGES OF LENSES OVER REFLECTOR ANTENNAS

Candidates for low-sidelobe antennas could be found among a number of well-established categories, such as reflector, lens, and array types. For reflectors and lenses, a primary feed illuminates a secondary, collimating aperture. For arrays, a primary feed is not used, but a feed transmission line network will be necessary to combine the powers from many antenna elements, and sidelobes are suppressed by weighting the elements' powers with a suitable aperture distribution. An asymmetric beam can be synthesized from an elliptical or rectangular array, again with suitably chosen elemental weights. Array antennas can be quite easily and cost-effectively fabricated using printed circuit technology, but for millimeter waves, the radiation efficiency would be expected to be rather low because of ohmic losses. Much better efficiency would be achieved using waveguide arrays, but at the cost of quite considerable fabrication complexity and cost. Array antennas can also have bandwidth limitation, caused by either the radiating element's bandwidth or that of the power combining network. Combining transmit and receive functions from a single array antenna could then be problematic if the respective frequencies for each are very much separated. In contrast, reflectors

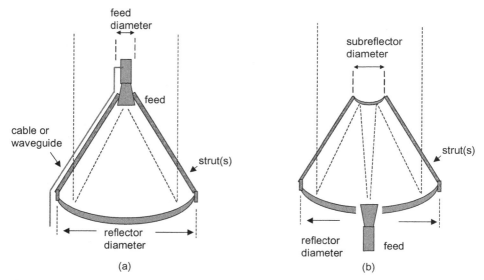

Figure 5.13. Non-offset-fed reflector geometries: (a) prime focus and (b) Cassegrain subreflector.

and lenses can more easily achieve wide bandwidths and very good polarization properties where a single antenna works in both transmit and receive modes: Typically, it is the primary feed in each case, along with polarizer and orthomode transducer in the RF chain, which allows for this.

5.3.1 Reflectors

For the reflector, some aperture blockage will tend to occur for either prime focus or subreflector geometry, illustrated in Figure 5.13. The latter includes Cassegrain and Gregorian types and, in each case, the subreflector introduces some blockage of the aperture, which can exceed that which would occur for the prime focus type. There might be several reasons for adopting a subreflector design over prime focus: Often, it is advantageous to eliminate the primary feed (and associated orthomode transducer and possibly transceiver electronics) from a location where they block the aperture. Also, mechanical considerations can make it preferable for these components to be housed behind the main reflector rather than supported on struts in front of the reflector aperture, and cable runs can also be reduced.

Nevertheless, a subreflector geometry is not always the best choice since the aperture blockage effect is not necessarily improved. This geometry is generally considered to be most worthwhile where very electrically large reflectors are to be used, and so the subreflector alone might be many wavelengths in diameter (overall antenna gain in the region from 40 to >50 dBi). Then, the cross section of the subreflector and struts becomes of less significance compared to the total aperture area.

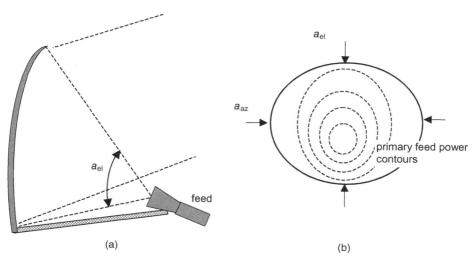

Figure 5.14. Offset-fed reflector geometry: (a) side view, (b) nonuniform illumination.

Of course, the above comments refer to non-offset-fed reflectors. Offset-feed types (either prime focus or subreflector) reduce or eliminate aperture blockage and very efficient and well-established types are encountered, from low-cost domestic satellite TV and very small aperture terminal (VSAT) applications through to space satellite payload antennas. However, while the overall optical efficiency of these types can be very high, this could be accompanied by some subtle disadvantages such as a loss of illumination symmetry, which can degrade cross-polar and sidelobe performance, properties well documented in the literature [13–15].

To show how this comes about, Figure 5.14 illustrates the way symmetry is lost when the reflector is offset fed. Consider a circular reflector where the feed is offset in elevation, as in Figure 5.14a. Seen from the feed's point of view, the reflector subtends a lesser angle a_{el} in elevation than in azimuth (a_{az}). While the feed need not be circular, this case has been chosen here for purposes of illustration. In any case, the primary feed may be chosen to have an elliptic radiation pattern so as to better illuminate the reflector, whatever shape that may take. The point, however, is that the reflector is being illuminated from an oblique angle, and so there is an asymmetry in the intensity of illumination along the elevation direction (Figure 5.14b). While these properties by no means present insurmountable obstacles to reflectors, they do render them a rather difficult starting point where we seek a modified, asymmetric beam with two carefully defined orthogonal beamwidths, very low sidelobes, and low cross polarization.

5.3.2 Lenses

It will now be apparent how, in contrast to reflectors (be they symmetrically or offset fed), lens antenna suffers no inherent aperture blockage. This can help suppress sidelobe

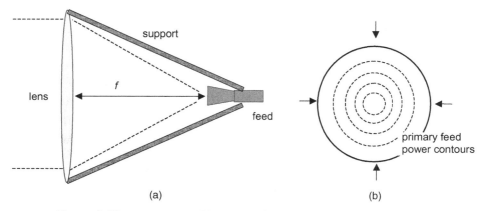

Figure 5.15. Lens antenna (a) lens and feed (b) symmetrical illumination.

radiation since there are no objects in front of the radiating aperture to cause unhelpful scattering. Furthermore, there is an inherent symmetry of illumination of the lens aperture by the primary feed (Fig. 5.15).

In addition, a lens surface can be modified to introduce the beam asymmetry desirable for the circular footprints described in the above section.

Where particularly low sidelobes are sought, this can be achieved by using a quite severe aperture taper. This approach leads to a reduced area efficiency (much of the aperture is very weakly illuminated), but this might not be the driving objective behind the antenna design. This has certainly been the case arising from the above discussion of antennas for HAPs, where sidelobes and controlled beam shapes are paramount. A severe aperture taper leads to a requirement for a relatively large primary feed aperture. This would, in turn, increase aperture blockage for a reflector antenna but presents no similar problem for a lens.

Figure 5.16 illustrates how a very low sidelobe radiation pattern can be produced by using a primary feed that is relatively large: Its diameter is $D1$, while that of the secondary aperture, the lens, is $D2$. In a practical design reported below, the primary feed is a medium gain antenna in its own right with gain of about 20 dBi. A lens is then used to increase the gain to around 30 dBi, a 10-fold increase in effective area. If a prime focus reflector was used in the same manner, 10% of the aperture would be blocked by the feed. The lens antenna does not exhibit this disadvantage.

5.3.3   Commercial Lens Antennas

Several commercial lens antennas are marketed with high efficiencies and low sidelobes. Some offerings for millimeter wave frequencies are available from Flann Microwave (see Fig. 5.17), Millitech, and Quinstar, among others.

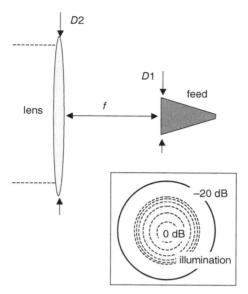

Figure 5.16. Severe aperture taper by primary feed.

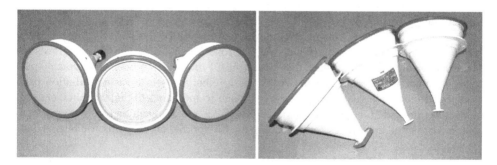

Figure 5.17. Commercial lens-corrected horn antennas.

The commercial antennas shown in Figure 5.17 have an aperture diameter of 160 mm and at 28 GHz offer E-plane and H-plane beamwidths, respectively, of about 5° and 6°, and a gain of 31 dBi. This, incidentally, would be about that required for a 6-km diameter cell toward the edge of the regions illustrated in Figure 5.3 with the caveat that the power footprints on the ground would tend to be elliptical and so coverage suboptimum.

A typical lens is shown in Figure 5.18. This has a diameter of 15 cm and has two refracting surfaces (one is shown). Both faces are also grooved to reduce gain ripple due to mismatch at the dielectric–air interface over a band of 28–32 GHz.

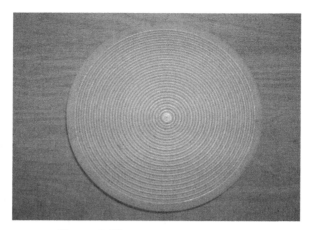

Figure 5.18. A grooved, circular lens.

5.4 DEVELOPMENT OF A SHAPED BEAM LOW-SIDELOBE LENS ANTENNA WITH ASYMMETRIC PATTERN

This section describes one approach to making a shaped beam lens antenna and is taken from a practical example [16].

A shaped beam lens antenna was developed as part of the EU *HeliNet* program (ca. 2003), which has been introduced in Sections 5.1 and 5.2. The objective was to construct experimental hardware that would fulfill the requirements of a typical spot beam antenna for HAP cellular networks and to then go on to measure radiation patterns. Section 5.2.1 presented a generic approach to modeling the interference caused by overlapping radiation patterns, and Sections 5.2.2 and 5.3 went on to show why lens antennas are good candidates for producing the low-profile and asymmetric beams that are required.

The target specification for this design exercise was HPBW of 5.0° in elevation and 9.4° in azimuth. The starting point was to implement in computer code an analytical model for generic lens radiation patterns after Silver [17] for a lens with one refracting surface which faces the primary feed. The aperture distribution is given by

$$A(r) = \sqrt{\frac{(n\cos\theta' - 1)^3}{f^2(n-1)^2(n-\cos\theta')}}F_p(\theta'), \tag{5.7}$$

where n is the refractive index of the lens material, r is the normalized aperture radius, and f is the focal length. The term inside the root is that inherent to the lens [17], while the term $F_p(\theta')$ is due to the primary feed. The first term comes about because of the curvature of the lens surface with respect to an incident spherical wave front as illustrated in Figure 5.19. Thus, for equal increments in solid angle $2\pi\Delta\theta'$, the constant incident energy is distributed across a greater surface area as θ' increases.

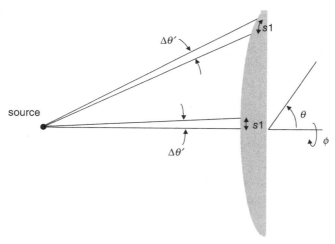

Figure 5.19. Aperture distribution across lens: reduction in projected power per unit solid angle.

The total aperture distribution is then the product of that inherent to the lens, and that of the primary feed. The secondary radiation pattern (the pattern of the lens with a given feed) may be computed from the transform of the aperture field Equation (5.7) to the far field. This can conveniently be evaluated by computer code by discretizing the aperture and using standard transform techniques, for example, discrete Fourier transform; this, in essence, invokes the method of physical optics (PO). A straightforward approach is to sample Equation (5.7) across a Cartesian grid in the plane of the aperture in steps of, say, one-quarter of a wavelength. Points outside of the circular aperture should be set to zero. This method was used to derive far-field patterns for the lens and feed combination. A corrugated feed used for this purpose is described below. In evaluating the far field of the lens illuminated by the corrugated horn, using computer code, it was found to be essential to take care that the phase component of the primary pattern was carefully preserved.

5.4.1 Primary Feed

In Equation (5.7), $F_p(\theta')$ is the far-field radiation pattern of the primary feed pattern as a function of angle θ' only, thus assuming a pattern that is symmetric; that is, it has no azimuth ϕ' dependency.

A corrugated horn was designed which could fulfill several purposes. First, this type of horn could meet the requirements of a medium gain spot beam antenna for cells very near the center of the coverage area (Fig. 5.3). The radiation pattern is circularly symmetric, has low sidelobes, very good polarization purity, and has very high efficiency. Second, these very properties make the same horn design well suited as a primary feed for a lens antenna for those more distant cells where greater directivity is required than that offered by the horn alone. For the first criterion (center cell spot beam), an HPBW of 17° was required. A horn was designed to meet this beamwidth at 28 GHz. A constant flare angle was chosen since this tends to exhibit lower sidelobes than the more compact profiled taper type.

From Reference 18, the radiation pattern is given by

$$F_p(\theta') = (1 + \cos\theta') \int_{r=0}^{r_1} J_0[kr\sin\theta'] J_0\left[2.405\frac{r}{r_1}\right]\exp\left(-j\frac{kr^2}{R}\right)\frac{r}{r_1}\,dr, \qquad (5.8)$$

where k is the wave number, r_1 is the horn aperture radius, and R is the flare angle of the conical taper (Fig. 5.20 and Table 5.1).

Two horns were fabricated by Thomas Keating Ltd in the United Kingdom using the electroforming process. Images of a horn are shown in Figure 5.21. The measured patterns for this horn at 28 GHz are shown in Figure 5.22a, and theory and measurement are compared in Figure 5.22b. The high symmetry of the patterns is evident.

5.4.2   Symmetric 5° Beamwidth Antenna

A conventional lens was fabricated from high-density polyethylene (HPDE) of diameter D 160 mm. This diameter, using a focal distance f of 192 mm ($f/D = 1.2$), rendered a 5° symmetric beam. The conventional lens curvature after Silver [17] is shown in Figure 5.23. The feed and lens coordinate systems are shown in Figure 5.24.

The beam symmetry is largely a consequence of the very high symmetry of the primary feed. So although the feed used to develop this lens antenna was of a quite expensive and bulky design (compared to a shorter, smooth wall conical horn), its very favorable radiation properties helped the later design of the asymmetric lens.

A measured radiation pattern for the antenna is shown in Figure 5.25. Although it is an H-plane pattern, the E-plane is almost indistinguishable because of the high beam symmetry.

5.4.3   Asymmetric Beam

Antenna pattern synthesis can be approached from two opposite directions. A formal, mathematical approach begins with the desired far-field radiation pattern and, through standard transform techniques, derives the required aperture distribution. This is well-trodden ground for manufacturers of communications satellite payload antennas, where there will be a requirement for a certain beam footprint as projected onto the surface of the earth. Such antennas are often reflectors with "shaped" surfaces (i.e., deviating from the classic paraboloid) or arrays, and tend to be high-cost items produced in low volumes. By analogy, we would expect similar methods to be applicable also to lenses, although that is not quite the approach taken here. For the purposes of achieving an asymmetric beam with low sidelobes, a second approach is discussed, which begins instead with the radiation pattern of a well-characterized corrugated horn-fed lens combination, and then seeks to perturb the aperture distribution so as to widen the

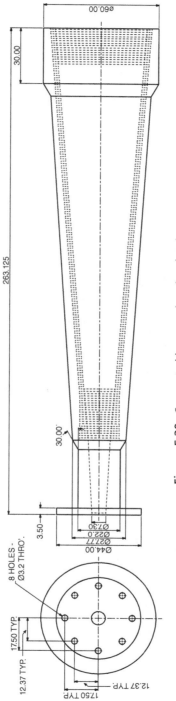

Figure 5.20. Corrugated horn: engineering drawings.

TABLE 5.1. Corrugated Horn Parameters

Aperture diameter	48 mm
Length	260 mm
Taper angle (half-cone)	5°
Input waveguide diameter	7.3 mm
HPBW	17°
Measured gain	21 dBi

Figure 5.21. Images of a corrugated horn.

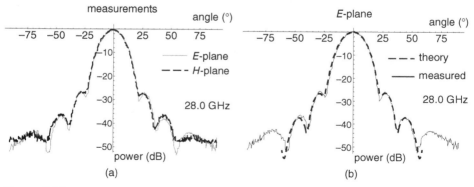

Figure 5.22. Corrugated horn (a) measured patterns and (b) measurement and theory compared, all at 28 GHz.

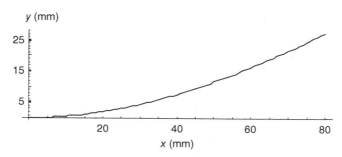

Figure 5.23. Lens profile: data provided to machinist.

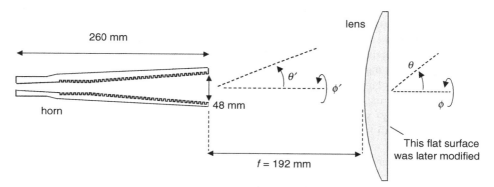

Figure 5.24. Lens and feed dimensions.

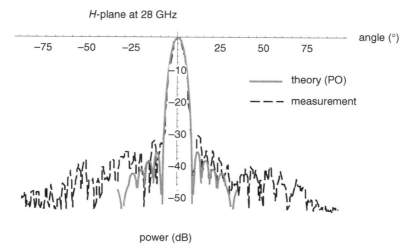

Figure 5.25. Radiation pattern of symmetric, corrugated horn-fed 160-mm HDPE lens at 28 GHz.

far-field pattern in one plane only. This is looking at the problem from the point of view of asking what modifications are necessary from a sound starting point (low-sidelobe lens antenna) rather than attempting to work directly from the required pattern backward toward the final antenna hardware. As such, it was a pragmatic design approach using readily available analysis and fabrication tools and one unlikely to arrive at a solution for the required electric field distribution across the lens aperture which could not be physically realized.

Another practical consideration was that of fabricating very small numbers of antennas (be they a lens, reflector, or array type) as individual items for purposes of experiment. Reflectors can be quite costly to produce in production runs of one. The means of fabrication available, as in any reasonably well-equipped machine shop, would be the use of a computer numerical control (CNC) machine to cut a billet of material with a curved surface, and in this respect, the use of plastics/polymers can be easier and less costly to work with than metallic billets.

The method then used was to investigate the effect of modifying the phase distribution of the lens aperture in one plane only (Fig. 5.26). The hypothesis to be tested was that this should change the beamwidth in one plane while leaving the radiation pattern in the orthogonal plane substantially unaltered. It should be accepted that this approach will serve only to increase the beamwidth in one plane, and so the aperture efficiency (or area efficiency) will inevitably be reduced because, for the same lens area, the modified beam will have reduced directivity.

Working still with the PO code, which transforms Equations (5.7) and (5.8) to far-field patterns, an additional routine multiplied Equation (5.7) by a phase distribution term of the form

$$\Delta\beta = F(x), \tag{5.9}$$

and in Figure 5.26,

$$x = r\cos(\phi - \phi'); \tag{5.10}$$

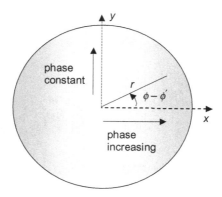

Figure 5.26. Modifying the lens aperture phase distribution in one plane only.

that is, phase shift $\Delta\beta$ is some function of x only in the plane of the aperture while remaining constant along the y-direction. From this approach, at least for small perturbations in phase distribution, the radiation pattern in the plane xz is expected to be directly affected, while the orthogonal beam in the plane yz should be only weakly affected. The phase term $\Delta\beta$ would then be physically realized from a thickness t of additional dielectric material according to

$$t = \frac{\lambda\Delta\beta}{2\pi(n-1)},\tag{5.11}$$

where n is the refractive index. To investigate the effect of a simple, linear phase slope across the lens aperture, the function

$$\beta = ax\tag{5.12}$$

may be used. Expressing a as a constant phase shift in degrees per unit wavelength, Equation (5.12) then introduces a uniformly linear phase shift across the aperture, which has the expected consequence of scanning the beam in the xz-plane. This effect is shown in Figure 5.27, where $a = 3°$ per wavelength and the pattern for the original lens antenna, with one nonrefracting (flat) surface, is also shown.

The computational method is illustrated graphically in Figures 5.28 and 5.29. In each, the lens aperture phase distribution is shown across both x- and y-axes in terms of the samples used for the PO transform. In Figure 5.28, the distribution is symmetric and caused by the primary feed alone. Also shown is the same phase distribution rendered along the x-axis only, the computed far field, and the curvature of the nonrefracting surface, this remaining unaltered. Then, Figure 5.29 shows how the introduction of the linear phase slope skews the aperture distribution: This squints the beam and may be physically realized introducing a dielectric "wedge" profile to the previously flat dielectric surface. The profile is derived from Equation (5.11) and assumes 28-GHz

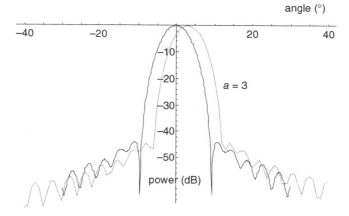

Figure 5.27. Beam squinting on the application of a linear aperture phase slope.

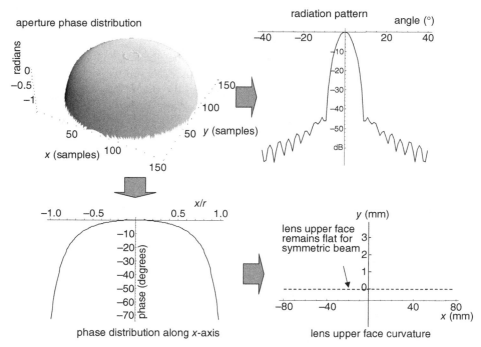

Figure 5.28. Lens aperture phase distribution due to primary feed only.

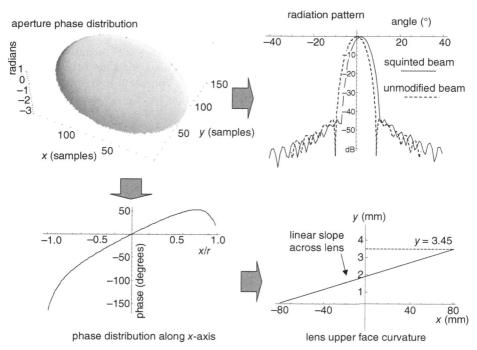

Figure 5.29. As Figure 5.28, multiplied by 2° per wavelength linear phase slope.

center frequency and $\varepsilon_r = 2.3$ (for polyethylene [PE]). The lens electrical diameter is very close to 15 wavelengths.

Thus far, Figures 5.28 and 5.29 have served to test the computational method and show how beam shape modification relates to a physical, easily manufacturable dielectric component. It remains to be found what kind of dielectric profile can bring about the beam asymmetry which is sought for the design example being considered. Clearly, a linear term will scan the beam rather than produce the desired increase in beamwidth. The next step (following the experimental methodology that was used) is to reverse the gradient of the linear phase slope at the center of the lens, according to

$$\Delta\beta = a|x|. \tag{5.13}$$

This phase distribution and its effect on the radiation pattern is shown in Figure 5.30, here using $a = 2°$ per wavelength. In a sense, this represents a small step in the right direction since there is evidence of the expected beam widening. However, the form of this is mostly unhelpful since the central part of the main lobe is largely unaffected, while the sidelobe levels are seriously degraded. (The nonmodified and modified beam cases have not been scaled for directivity.)

The effect of a more extreme phase variation is shown in Figure 5.31, here showing only the phase distribution along the x-axis (Figure 5.31a) and the radiation pattern

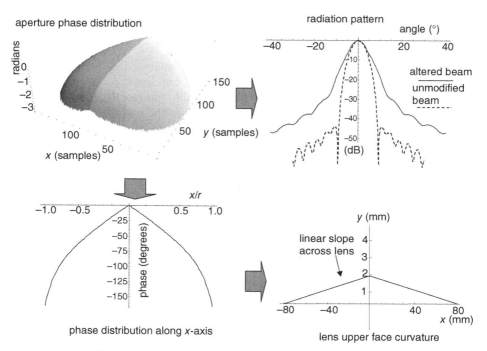

Figure 5.30. Effect of moderate linear phase term in |x|.

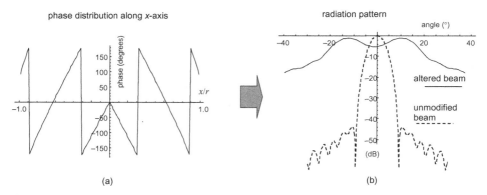

Figure 5.31. Effect of excessive linear phase term in |x|.

(Figure 5.31b). The former wraps at 2π and so could be realized physically with a faceted (or zoned) dielectric surface. The effect on the radiation pattern is bifurcation of the main lobe and along with this would be a severe reduction in directivity. This example is not useful for the present design exercise but is shown for academic interest.

Progress can be achieved by increasing the number of degrees of freedom in the curvature of the second refracting surface of the lens, that is, by using a polynomial rather than a linear phase distribution. This can have the form

$$\Delta\beta = a|x| + b|x|^2 + c|x|^3 + \ldots. \tag{5.14}$$

A polynomial of this type may also be conveniently combined with an optimizing routine in an attempt to iteratively approach a required beam shape. A convenient model for the main lobe is that using Equation (5.2) since this can rapidly generate a main lobe from a single input parameter, namely, HPBW. In the example presented, this method was applied to find a polynomial that gave a best fit to the target HPBW of 9.4° in the plane of the widened beam.

Following several iterations with the PO code where the phase adjusting polynomial coefficients were varied, the terms arrived at were $a = 0$, $b = 1$, and $c = -0.07$ (degree per wavelength) in Equation (5.14). This yields the theoretical radiation patterns shown in Figure 5.32. Also shown are the "target" main lobe patterns, which show very close agreement to the PO-generated main lobes. The phase profile is shown in Figure 5.33a and the dielectric lens profile in (Figure 5.33b). This profile shape was then machined onto the second (or outer) face of the lens, that is, the previously flat surface for the case of the simple lens and symmetric beam. (Note that Fig. 5.33b is *not* a scale drawing of the lens profile.)

5.4.4 Measurements

The fabricated PE lens (Fig. 5.34) was illuminated with the corrugated horn primary feed illustrated in Figures 5.20 and 5.21. The experimental arrangement is shown in

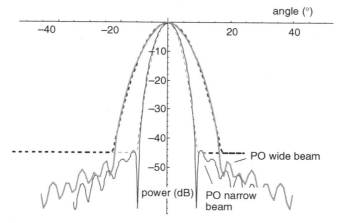

Figure 5.32. PO (theory) for wide and narrow beams and target curves (dashed).

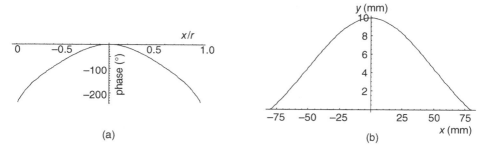

(a)   (b)

Figure 5.33. Phase distribution after optimization (a) and dielectric profile (b).

Figure 5.34. Photo of profiled polyethylene (PE) lens.

Figure 5.35. Photo of lens and feed in a measurement chamber.

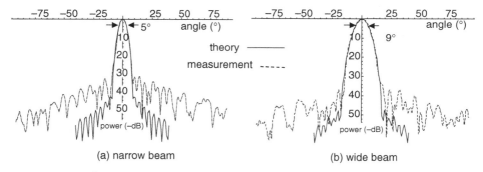

Figure 5.36. Measured asymmetric beam radiation patterns.

Figure 5.35 (and, incidentally, is that described in Fig. 1.28 and the description that follows in Chapter 1).

To sum up, the lens had two diffracting surfaces: (i) a surface facing the feed and having the classic hyperbolic profile (Fig. 5.23) which collimates the primary field, and (ii) a second surface, profiled in one axis as according to Figure 5.33 to increase the beamwidth in one plane. The profile of the second surface had been optimized to find the best fit to a target beam pattern, as described above. The measured radiation patterns are shown in Figure 5.36 for the narrow and wide beamwidth cases. These are not referred to here as either H-plane or E-plane because the primary feed pattern is very symmetric and so rotation of the primary feed about its longitudinal axis will not affect these patterns.

The measured patterns for the asymmetric beam are shown again in Figure 5.37, here along with computed average sidelobe floors. These equivalent sidelobe levels have been computed as the mean power between nominal points close to the first null,

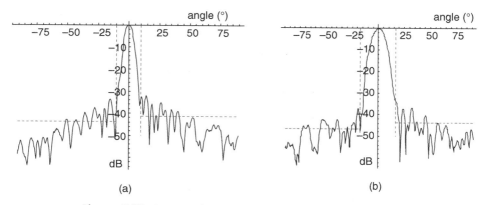

Figure 5.37. Measured patterns showing mean sidelobe levels.

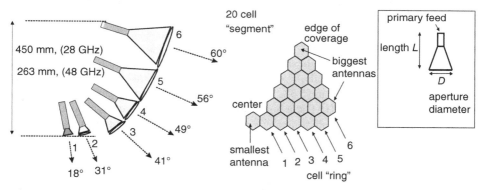

Figure 5.38. Antenna pointing angles where six concentric cells are illuminated by a HAP.

and ±90°. Accordingly, the mean levels are slightly different on the negative angle axis compared to the positive. Representing the sidelobes as flat floors is of course simplistic because sidelobes tend to decay with increasing angle, as is apparent in Figure 5.37. However, flat equivalent levels have been shown because for networks of many overlapping antenna footprints, as encountered in the cellular networks discussed above and which lead to the design of this antenna, these can be a useful and relevant metric. The mean levels (considering both sides of the pattern) are −42 and −44 dB, respectively, with respect to peak, boresight gain, for the narrow and wide beamwidth lobes.

5.5 LENS ANTENNA PAYLOAD MODEL

As part of the European *HeliNet* project, which has been introduced above, some thought was put into how a group of "dedicated" antennas (one per cell) might be organized on a HAP payload. This is reported in Reference 19, from which Figures 5.38 and 5.39 are taken.

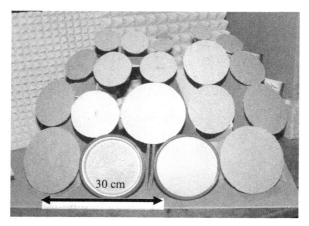

Figure 5.39. Lens antennas: a model for part of a flight payload.

From the symmetry of the hexagonal cell layout of Figure 5.3, it is only necessary to consider one-sixth of the cells. These are shown as a "segment" of the coverage area, comprising 20 cells, in Figure 5.38. Here, the smallest antennas point to the central cells: These cells view the HAP at a high elevation angle. For more distant cells, elevation decreases (or the beam pointing angle with respect to the zenith increases): smaller beamwidths and therefore larger antennas are required.

A way of organizing the antennas is shown as a full-scale model in Figure 5.39, where a carrier frequency of 28 GHz has been assumed. Two real antennas of the type shown in Figure 5.17 are used. The remaining "apertures" are dummies but have the correct scaling in accordance with the dimensions shown in Figure 5.38.

This model is also "upside-down"—the apertures face upward rather than downward. The purpose of this model was to gain at least an approximate appreciation of the dimensions required for a dedicated beam payload since this, in turn, informed the wider aspects of the *HeliNet* project. Six antenna groups would be needed in total. It has also been proposed that these groups could be steered independently to accommodate changes in platform altitude and the displacements that would inevitably occur in the winds of the stratosphere.

5.6 MULTIFEED LENS

Another useful property of lenses for multibeam applications (e.g., HAP payloads) is their relative intolerance to being fed off-axis, particularly where the focal length/diameter ratio is long (e.g., >1), thus lending themselves to multibeam deployments, as illustrated in Figure 5.40. This could reduce the total number of lens apertures required for a payload of the type discussed above and illustrated in Figures 5.38 and 5.39.

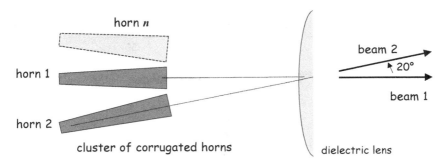

Figure 5.40. Multifeed lens antenna.

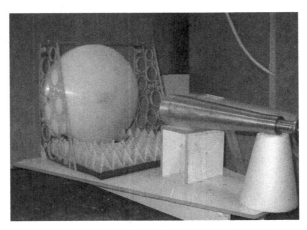

Figure 5.41. Experimental low-sidelobe, multiple-beam lens antenna.

This geometry was explored experimentally as part of the EU *HeliNet* project, although the results were not published beyond the project's internal deliverables. The antenna shown in Figure 5.41 uses a conventional dielectric lens with a radius of 250 mm and fed by a pair of corrugated horns at 28 GHz. This gives rise to a highly underilluminated aperture, and thus aperture efficiency is sacrificed to achieve very low sidelobes. The secondary radiation pattern (far field of lens) is circularly symmetric, and so this version does not produce circular cells at low elevation angles (Fig. 5.42)

The scan angle is up to 20° and nine feeds (and hence beams) can be accommodated. The sidelobe floor here is at −40 to −60 dB below peak gain. The logical extension of the multibeam lens concept is the spherical class of lenses, which includes Luneburg lenses, discussed in the following section and also in the later chapters.

It is worth reiterating that, where equal-size cells are needed, the antenna payload would employ a variety of different size antennas where the near-cells (at high elevation) require the smallest aperture, and so horn-type, symmetric beam antennas would

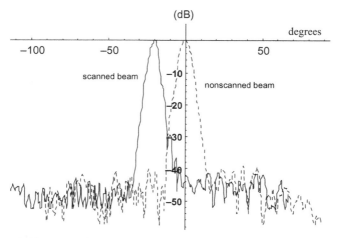

Figure 5.42. Measured radiation patterns, ultra-low-sidelobe lens antenna.

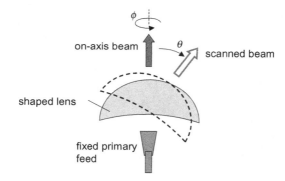

Figure 5.43. Beam scanning where primary feed is fixed.

be a pragmatic choice. For distant cells (at medium to low elevation), larger antennas are needed and asymmetric lens types are useful here. These considerations were studied in some detail under the *HeliNet* project and are summarized in Reference 19.

Off-axis scanning in a lens antenna may also be exploited to steer a single beam. In Reference 20, the lens curvature was optimized to provide scanning to ±45° at 60 GHz and for an antenna gain in the region 21 dBi. The lens was illuminated by a fixed conical horn-type primary feed (Fig. 5.43). Lenses with both a single refracting surface and with two refracting surfaces were investigated, the latter offering more scope for optimization. In a demonstration, PE was used for the lens material. A cited advantage is the ability to scan the antenna beam while leaving the feed fixed and so eliminating rotating waveguide joints, which tend to be costly or troublesome at these frequencies.

In Reference 20 and Figure 5.43, a simple gimbal supports scanning in the two axes: θ increases as the lens is tilted with respect to the on-axis beam while rotation ϕ

is around the feed axis. θ introduces a scanning loss, designed here to be minimized over $\pm 45°$, while from symmetry, ϕ is any arbitrary angle between $\pm 180°$ and does not bring about scan loss.

5.7 MULTIPLE BEAM SPHERICAL LENS ANTENNAS FOR HAP PAYLOAD

The introduction to this chapter discussed how HAPs could exploit millimeter wave spectrum using multicell architectures and that the capacity of a communications network would be dominated by the payload antennas which serve the cells. The best coverage would be expected when each antenna pattern is tailored to the footprint of its respective cell on the ground. This requires asymmetric beam antennas and the lens antenna described in Section 5.4 has been put forward as a promising candidate and also as a general approach for controlling the antenna's two orthogonal beamwidths. This has been called the dedicated beam model for HAP communications and presents an ideal case from a capacity and interference point of view, but also a rather bulky antenna payload.

In contrast, multibeam antennas could reduce the size and mass of the antenna payload by reducing the number of antenna apertures. Such antennas would use multiple primary feeds and share the main radiating aperture. An example has been discussed in Section 5.6, where multiple feeds were used with a fairly conventional focusing lens. Of course, this tends to reduce the designer's freedom in fine-tuning the radiation pattern for each beam. Taking this philosophy to a more extreme case, spherical lenses allow for a very high reuse of the aperture (the lens) because many primary feeds can be used. The electromagnetic properties of spherical lenses are discussed in detail in Chapter 6, but for present purposes, we present a system-level study of how they might be used for HAPs.

A classic type of spherical lens is the Luneburg lens, which has been encountered in the introductory chapter. Variants include constant-index and stepped-index techniques.

An aperture distribution for the Luneburg lens is given by Morgan [21]. This is a convenient starting point for the present analysis because the primary feed pattern may readily be taken into account: From this, the far-field radiation pattern of the lens can be computed. The geometry is illustrated in Figure 5.44.

Using again the computationally convenient model for the primary feed power pattern P_{PF} as a function of angle θ',

$$P_{PF} = \cos[\theta']^n, \qquad (5.15)$$

where

$$\theta' = \arcsin\frac{r}{f}, \qquad (5.16)$$

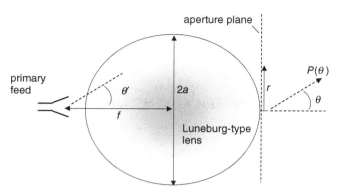

Figure 5.44. Morgan's geometry [21] for generalized feed and Luneburg lens.

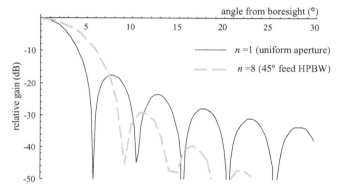

Figure 5.45. Theoretical radiation patterns for Luneburg lens with a 10-wavelength diameter.

we then obtain the radial power distribution $P(r)$ in the aperture plane:

$$P(r) = \frac{\cos\left[\arcsin \dfrac{r}{f}\right]^{n}}{f^{2}\sqrt{f^{2} - r^{2}}}. \tag{5.17}$$

The far-field power pattern $P(\theta)$ is then obtained by the transform of Equation (5.17) to the far field. In a special case where $n = 1$, the r dependence of $P(r)$ disappears and we have a uniform aperture. For a reduced beamwidth primary pattern (increasing n), the lens is less intensely illuminated at its edges, leading to lower sidelobes and reduced directivity. To then recover the directivity required for a given cell on the ground, the lens diameter would need to be increased. Equation (5.17) thus offers a simple recipe for trading lens size, feed pattern, and sidelobe levels. A similar analysis is explored in Reference 22. Using this approach, radiation patterns are shown in Figure 5.45, where the lens diameter is 10 wavelengths.

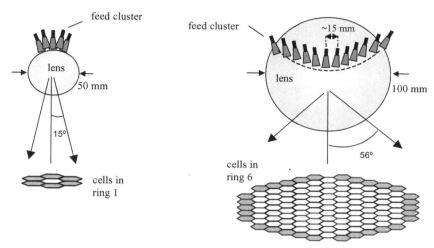

Figure 5.46. Concept for antenna payload: multibeam Luneburg lens for each cell group.

In the layout of 121 cells shown in Figure 5.3, the cells are arranged in six concentric hexagonal rings. Since we are now proposing shared apertures for cells within each ring, the azimuth and elevation beamwidths cannot be chosen independently. Instead, we can choose an antenna beamwidth based on the *mean* required value. This leads to noncircular cellular power footprints but allows the use of a single spherical lens aperture for each ring and thus a much more compact antenna payload. Parts of such a payload are illustrated in Figure 5.46, which shows how the innermost group of cells is served by the smallest spherical lens with a cluster of appropriately spaced feeds, and the outermost group of lens is similarly served by the largest spherical lens. The lens diameters indicated are approximations based on uniform aperture illumination and a 28-GHz carrier frequency and thus serve as a reasonable estimate for the minimum antenna dimensions.

Figure 5.47 shows power contours for a cochannel cell group where Figure 5.47a is for dedicated beams, which are asymmetric, and Figure 5.47b is for circular symmetric beams, which could be generated using multibeam spherical lenses.

The lens radiation pattern that has been used in Figure 5.47b is the uniform aperture model and hence the term "minimum size"—a smaller aperture cannot yield the required beamwidths. However, this radiation model also has the worst sidelobes compared to an aperture with the tapered illumination, which would be brought about by a more directive primary feed, as shown in Figure 5.45. Unsurprisingly the sidelobe properties in Figure 5.47b give rise to inferior cochannel interference than the dedicated beam model of Figure 5.47a. To redress this disadvantage, the lens diameter may be increased and used with a directive primary feed to reduce sidelobe levels. This gives rise to an increase in the bulk of the antenna payload as a whole. A full analysis is found in Reference 23. To summarize this, the diameter of each spherical lens serving a cell group would need to be increased by a factor of about 1.2 (compared to the smallest

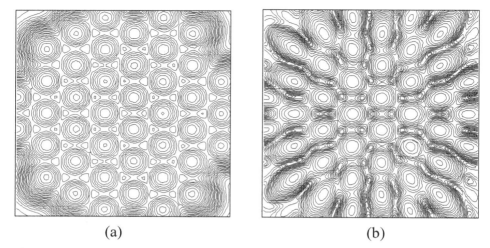

(a)   (b)

Figure 5.47. Power contours for one of three channels: (a) dedicated beams and (b) multiple beams, minimum-size spherical lenses (contour spacing is 1.5 dB).

case, the uniform aperture) and used with 60° HPBW feeds to achieve comparable CIR performance to a dedicated beam model. To illustrate the effects at work, the uniform aperture first sidelobe level is at −17 dB (well-trodden ground indeed), while for the 60° HPBW feed and the transform of Equation (5.17), the first sidelobe is suppressed to −24 dB. This sidelobe reduction pulls down the average cochannel interference but at the expense of antenna size.

Thornton [23] goes on to estimate how the mass of these two contrasting antenna payloads would differ, offering figures for dielectric material alone of the order of 30 kg for 121 dedicated antennas, but reducing to 2 or 3 kg for multibeam spherical lenses. Such antennas have not yet been deployed in practice on HAPs, but the advantages of lens antennas for this still emerging area should be apparent. Cellular communications from HAPs is discussed further in Reference 24, alternative antenna configuration such as steerable arrays in Reference 25, and the radio regulatory environment in Reference 26.

REFERENCES

[1] G. M. Djuknic, J. Freidenfelds, and Y. Okunev, "Establishing Wireless Communications Services via High-Altitude Aeronautical Platforms: A Concept Whose Time Has Come?" IEEE Communications Magazine, September 1997, pp. 128–135.

[2] N. J. Collela, J. N. Martin, and I. F. Akyildiz, "The HALO Network," IEEE Communications Magazine Vol. 38, June 2000, pp. 142–148.

[3] T. C. Tozer and D. Grace, "High-Altitude Platforms for Wireless Communications," IEE Electronics and Communications Engineering Journal Vol. 13, Issue 3 June 2001, pp. 127–137.

[4] J. Thornton, D. Grace, M. H. Capstick, and T. C. Tozer, "Optimizing an Array of Antennas for Cellular Coverage from a High Altitude Platform," IEEE Transactions on Wireless Communications Vol. 13, Issue 3 No. 3, May 2003, pp. 484–492.

[5] J. Thornton, D. Grace, C. Spillard, T. Konefal, and T. C. Tozer, "Broadband Communications from a High Altitude Platform—The European *HeliNet* Programme," IEE Electronics and Communications Engineering Journal Vol. 13, Issue 3, June 2001, pp. 138–144.

[6] D. Grace, M. H. Capstick, M. Mohorcic, J. Horwath, M. B. Pallavicini, and M. Fitch, "Integrating Users into the Wider Broadband Network via High Altitude Platforms," IEEE Wireless Communications Vol. 12, No. 5, 2005, pp. 98–105.

[7] D. A. J. Pearce and D. Grace, "Optimum Antenna Configurations for Millimeter-Wave Communications from High-Altitude Platforms," IET Communications Vol. 1, No. 3, 2007, pp. 359–364.

[8] J. Holis and P. Pechac, "Elevation Dependent Shadowing Model for Mobile Communications via High Altitude Platforms in Built-up Areas," IEEE Transactions on Antennas and Propagation Vol. 56, No. 4, 2008, pp. 1078–1084.

[9] H. Taub and D. L. Schilling, Principles of Communication Systems, 4th ed., McGraw-Hill, New York, 1989, Chapter 16.

[10] R. J. Bates, Broadband Telecommunications Handbook, 2nd ed., McGraw-Hill, New York, 2002, Chapter 21.

[11] S. Haykin, Communication Systems, John Wiley & Sons Inc., New York, 2001, Chapter 7.

[12] J. Thornton and D. Grace, "Effect of Lateral Displacement of a High Altitude Platform on Cellular Interference and Handover," IEEE Transaction on Wireless Communications Vol. 4, No. 4, July 2005, pp. 1483–1490.

[13] C. A. Balanis, Antenna Theory, Analysis and Design, 2nd ed., Wiley, Hoboken, NJ, 1997.

[14] J. D. Kraus, Antennas, 2nd ed., McGraw-Hill, New York, 1988, pp. 745–749.

[15] A. W. Rudge, K. Milne, A. D. Olver, and P. Knight, The Handbook of Antenna Design, IEE Electromagnetics Waves, Series 15. Peter Peregrinus Ltd, London, 1982, pp. 526–543.

[16] J. Thornton, "A Low Sidelobe Asymmetric Beam Antenna for High Altitude Platform Communications," IEEE Microwave and Wireless Components Letters Vol. 14, No. 2, February 2004, pp. 59–61.

[17] S. Silver, Microwave Antenna Theory and Design, Peter Peregrinus Ltd, London, 1984, p. 391.

[18] A. D. Olver, P. J. B. Clarricoats, A. A. Kishk, and L. Shafai, Microwave Horns and Feeds, IEE Press, London, 1994.

[19] J. Thornton, "Antenna Technologies for Communications Services from Stratospheric Platforms," 4th International Airship Convention and Exhibition, Cambridge, UK, July 28–31, 2002.

[20] J. R. Costa, E. B. Lima, and C. A. Fernandes, "Compact Beam-Steerable Lens Antenna for 60 GHz Wireless Communications," IEEE Transactions on Antennas and Propagation Vol. 57, No. 10, October 2009, pp. 2926–2933.

[21] S. P. Morgan, "General Solution of the Luneberg Lens Problem," Journal of Applied Physics Vol. 29, No. 9, September 1958, pp. 1358–1368.

[22] J. Thornton and D. Grace, "Effect of Antenna Aperture Field on Co-channel Interference, Capacity and Payload Mass in High Altitude Platform Communications," ETRI Journal Vol. 26, No. 5, October 2004, pp. 467–474.

[23] J. Thornton, "Properties of Spherical Lens Antennas for High Altitude Platform Communications," 6th European Workshop on Mobile/Personal Satcoms & 2nd Advanced Satellite Mobile Systems (EMPS & ASMS). ESTEC, European Space Agency, September 21–22, 2004.

[24] J. Thornton, D. A. J. Pearce, D. Grace, M. Oodo, K. Katzis, and T. C. Tozer, "Effect of Antenna Beam Pattern and Layout on Cellular Performance in High Altitude Platform Communications," International Journal of Wireless Personal Communications Vol. 35, No. 1–2, October 2005, pp. 35–51.

[25] G. White, Y. Zakharov, and J. Thornton, "Array Topologies for High-Altitude Platform Smart Antennas," 8th International Symposium on Wireless Personal Multimedia Communications (WPMC 2005), Aalborg, Denmark, September 17–22, 2005.

[26] J. Thornton and T. Tozer, "Adaptation of ITU Recommended Antenna Pattern for Prediction of Multi-Cell Co-channel Interference in High Altitude Platform Communications in the 47/48 GHz Band," The Seventh International Symposium on Wireless Personal Multimedia Communications (WPMC) 2004, Abano Terme, Italy, September 12–15, 2004.

6

SPHERICAL LENS ANTENNAS

John Thornton

This chapter discusses the general class of spherical lenses and the chief variants of constant-index and variable-index lenses. Following an introduction to the basic properties, a short history and review of literature is given. Then, analytical methods are discussed, beginning with ray tracing and then the more powerful spherical wave expansion (SWE) method. Some results from commercial solvers are presented. The constant-index lens is considered in more detail. Practical stepped-index hemisphere lenses using few dielectric layers are then discussed in Chapter 7.

6.1 INTRODUCTION

Spherical lens antennas possess a unique property: the ability to produce multiple beams by the use of multiple primary feeds that share the lens aperture. The basic geometry of the multibeam spherical lens antenna is shown in Figure 6.1.

Before reviewing in more detail the properties of this class of antenna, including its chief variants, it is worth discussing the essential features of the simple geometry

Modern Lens Antennas for Communications Engineering, First Edition. John Thornton and Kao-Cheng Huang.
© 2013 Institute of Electrical and Electronics Engineers. Published 2013 by John Wiley & Sons, Inc.

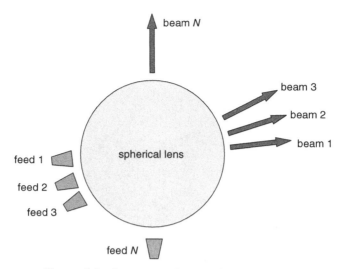

Figure 6.1. Illustration of spherical lens geometry.

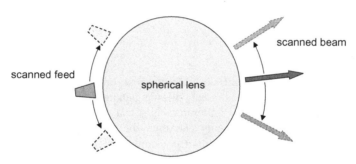

Figure 6.2. Spherical lens as a scanning antenna.

presented. In common with both the reflector antenna and the single-feed lens, the spherical lens antenna exhibits a primary aperture (the feed) and a secondary aperture (the lens). The secondary aperture is typically electrically large, and so the antenna has moderate to high directivity. It is the spherical symmetry of the lens that allows for the use of multiple primary feeds to generate beams that do not exhibit scanning loss. This is not to say that the efficiency of the spherical lens antenna is necessarily very high (although it can be) nor that other disadvantages are not encountered. Rather, it is the property of symmetry and thus uniformity of each beam, be they from multiple feeds as illustrated or from a scanned feed, which is of particular merit. Thus, either a multiple beam, a switched beam, or a scanning beam antenna (Fig. 6.2) may be produced, each of these cases being closely related variants.

The advantages for beam scanning compared to a multifeed reflector are apparent if we consider the properties of the latter, as illustrated in Figure 6.3. Here, multiple

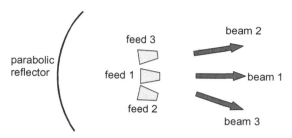

Figure 6.3. Multifeed reflector antenna.

beams may be generated by placing additional feeds in close proximity to the focal point of the reflector. Now, if we suppose that "feed 1" is at the focal point, then "beam 1" will exhibit high aperture efficiency. However, its near neighbors (feeds 2 and 3) are compromised by being placed off-axis, that is, at nonfocal locations, and so will exhibit scanning loss. This loss may be tolerable, perhaps a few decibels for closely spaced beams, but quickly becomes catastrophic as wider scan angles are sought.

An additional disadvantage is the increased level of aperture blockage by the region of space in front of the reflector now being occupied by primary feeds. This can be mitigated by using an offset-fed geometry, but here other compromises will be encountered such as nonuniform aperture illumination and a tendency for cross-polar levels to rise. Offset-fed reflector geometries are of course very highly developed and represents a successful and economic class of antenna often encountered in satellite broadcasting or communications, but multifeed types can still exhibit the above cited compromises.

A closely related antenna type is the toroidal reflector. Here, a reflector is "stretched" along one axis (e.g., the azimuth axis for a multiple beam satellite television antenna), and so the toroid resembles a number of overlapping reflectors. As such, the aperture is not fully shared and its area is also increased without a corresponding increase in directivity. Nevertheless, the toroidal reflector type is quite effective for scan angles up to about 40° and is encountered as a commercial offering for multiple satellite television reception.

A tracking or steerable antenna may be produced by taking a reflector antenna in its entirety and adding a motorized platform to steer in one or more axes, as illustrated in Figure 6.4. This is a very practical and pragmatic solution for many applications, for example, vehicular communications, surveillance, or radar, and need not be explored in great detail here. The method is noted by way of comparison with the spherical lens, which can similarly produce a steerable beam but by movement of the feed only (the lens may remain fixed to a vehicle's structure). This offers the advantage of moving a relatively small mass; that is, the primary feed and can also reduce wind loading. Additionally, the spherical lens may be used with multiple independently steered feeds. Some practical developments in this area will be explored in the final chapter.

The ray focusing properties of reflector and lens antennas are illustrated in Figure 6.5. Here, Figure 6.5c shows only the paraxial rays since a constant-index lens is

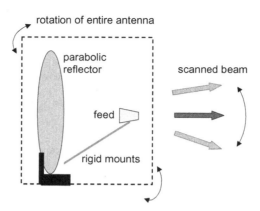

Figure 6.4. Mechanically scanned reflector.

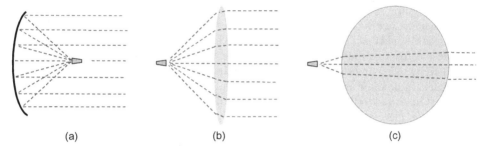

Figure 6.5. Ray focusing illustration of (a) a prime focus reflector, (b) a conventional focusing lens, and (c) a constant-index spherical lens.

considered and thus not all rays are properly focused. Later, ray-tracing models for constant and stepped-index lenses will be shown in better detail along with a more complete analysis of their focusing properties. The purpose of Figures 6.5 and 6.6 is to illustrate how beam scanning may be achieved in the three different geometries shown and how, for the prime focus reflector and conventional focusing lens, it is necessary to rotate the primary feed away from the optical axis, that is, displaced from the focal point labeled f in Figure 6.6.

Also in Figure 6.6, for purposes of clarity, only exit rays are shown, and for the spherical lens (Fig. 6.6c), the focus is characterized by a spherical surface rather than the focal points in Figure 6.6a,b.

Figure 6.7 illustrates scanning loss trends for the antenna types shown in Figure 6.6. The axis for scan angle is not scaled for any particular units because the figure is intended only to show a trend, and in particular that the spherical lens does not exhibit scan loss. For the other two cases, the rate of falloff of the loss curve will depend on many factors including ratio of the focal length to the aperture diameter.

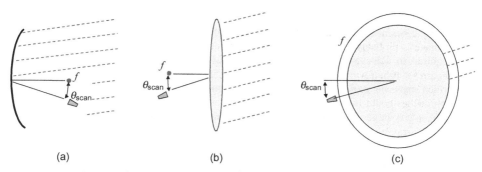

(a)   (b)   (c)

Figure 6.6. Defocusing in (a) a prime focus reflector, (b) a conventional focusing lens, and (c) a constant-index spherical lens.

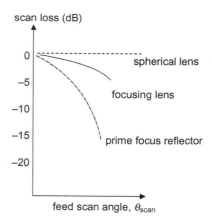

Figure 6.7. Scanning loss trends.

The introduction would not be complete without visiting briefly the properties of phased-array antennas by way of comparison with the spherical lens. The phased array uses electronic means to scan a beam in the desired direction without recourse to moving parts. The literature is extensive: The reader is directed to Reference 1 for a brief introduction and Reference 2 for a more comprehensive review. However, the views expressed in this chapter have been shaped to some extent by research into antennas for communications to both high-altitude platforms and satellites, topics visited in earlier chapters and where a requirement for very wide scan angles was encountered. Borrowing from Reference 3, written in 2006 and summarizing a critique of scanning antennas in the context of HAP research supported by the European Commission:

> The term "smart antenna" is often used in this context. In terms of complexity and ambition, this solution is at the opposite extreme compared to the mechanically steered dish. Some recent programs in this field are reviewed. The main advantages include:

conformality of the array, electronic beam scanning, rapid scanning over wide angles, multiple beam forming, free space combining of transmitted power. The main disadvantage is probably the very large number of antenna and active elements which are required to achieve a high gain aperture. Added to this, a conventional approach requires sub-half wavelength element spacing. This leads to very complex RF circuitry which may be close to being physically non-implementable. Less conventional approaches could include the use of widely spaced elements, sub-arrays etc. In any case, a very large number of elements are required, which will be extremely costly. A planar array also exhibits scanning loss which inevitably reduces the aperture gain where it is most needed.

The reference here to scanning loss should be taken in the context of a requirement to achieve very wide scan angles—not a strong point of planar arrays. That critique went on to conclude

> Having reviewed a number of techniques which attempt to reduce the complexities of array topologies for smart antennas, it was felt that none offer the required functionality in terms of efficiency, scan performance and cost. Often, when a technique is put forward to mitigate some disadvantage (e.g. use of sub-arrays to reduce number of active components), some other disadvantage is made apparent (e.g. reduced scan angle or appearance of grating lobes).

To follow this up with some further reflections on that earlier program of work, it might be said that array techniques were of particular interest and popularity among academics and theorists, but somewhat less attractive when viewed by the practitioners of experiment and implementation, particularly in the context of scanning antennas for higher microwave and millimeter wave bands. This note of pragmatism is likely to echo throughout this chapter.

6.2 SPHERICAL LENS OVERVIEW

Two main classes of spherical lens are encountered in literature dating back to the 1940s. These two classes are loosely the constant-index lens and the variable-index or Luneburg lens. The former class is entirely straightforward since the lens comprises a sphere of homogeneous dielectric material, while the latter class comprises several variants, and care should be taken with terminology. The Luneburg lens [4] is a structure where dielectric constant ε_r varies as a function of radius r:

$$\varepsilon_r = 2 - \left(\frac{r}{R}\right)^2, \tag{6.1}$$

where R is the lens radius.

In theory, the lens efficiency can approach unity since, from a ray-tracing perspective, all the rays in a plane wave incident on the lens aperture arrive at a single focal point, which is at the lens edge, $r = R$. Alternatively, all rays emanating from a point source at the lens edge are focused at infinity. We should now be mindful of two variants that arise for quite different reasons:

(i)   a spherical lens with two arbitrary foci, that is, not necessarily at $r = R$ and $r = \infty$, and

(ii) a layered or stratified structure that is designed to approximate a continuous radial variation in dielectric constant.

The former (i) lens type was generalized by Morgan [5]: Its chief utilities include realizing a focal distance that can lie a distance from the lens edge, and a greater freedom over DC values. A Luneburg lens is thus a subclass of the general formulation from Morgan where one focus is at $r = R$ and the other at infinity, and also where the index must be unity at the lens outer surface. Another property of the Luneburg formula is that the maximum index required, which occurs at the lens center, is minimized. This tends to lead to the lowest DC values (i.e., between unity and 2) across the lens and thus also tends to minimize lens mass compared to those relying on formulae that utilize higher constants, albeit where this might entail a smaller variation in constant. Morgan also cited solutions by Brown and Gutman, where one focus may lie inside the lens surface, according to

$$ \varepsilon_{\mathrm{r}} = \frac{1 + \left(\dfrac{r_1}{R}\right)^2 - \left(\dfrac{r}{R}\right)^2}{\left(\dfrac{r_1}{R}\right)^2}, \tag{6.2} $$

where the second focus is at infinity and r_1/R is the normalized radius of the first, internal focus. Clearly, Equation (6.2) reduces to Luneburg's solution (Eq. 6.1) if $r_1 = R$.

The latter (ii) stratified type is a consequence of fabrication issues: The "ideal" Luneburg equation is extremely difficult to accurately realize in a fabricated structure, and so the layered or stepped-index approach is typically encountered in practice. Designs are often of a uniform type where shell thickness is constant; that is, shell radius is incremented in uniform steps. This design might often be referred to as a uniform stepped-index lens or less accurately as a Luneburg lens. Figure 6.8 illustrates how concentric shells of dielectric material are typically used in practice, and Figure 6.9 shows a 45-cm-diameter hemisphere lens manufactured according to this principle.

A third variant arises from the second, where the shell steps are not uniform [6] and/or there is no attempt to approximate the Luneburg radial distribution. This class should in general be called the non-uniform stepped-index lens. Several practical designs have been reported in literature, such as Cornbleet's [7] 1965 report of a three-layer lens offering 25.5 dBi at X band [6]. reported how a two-layer lens can offer high efficiency in theory even for an electrically large lens (30-wavelength radius) but reported a solution that required very fine tuning of DC values to three decimal places and that is unlikely to be feasible for manufacture with practical materials. Thornton [8] reported in 2005 a pragmatic approach to a two-layer non-uniform stepped-index

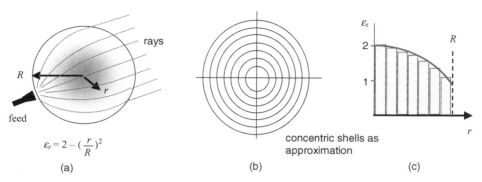

Figure 6.8. Luneburg lens and discretized approximation for practical fabrication: (a) rays in "ideal" Luneburg lens, (b) concentric layers and (c) radial steps as approximation.

Figure 6.9. An eight-layer stepped-index lens.

design using common materials and which was soon thereafter demonstrated in practice: This is described in the Chapter 7.

A "family tree" of spherical lens types is shown in Figure 6.10.

A body of work was published by Sanford in the mid-1990s, including a full analysis of the scattering matrix of stratified spherical lenses [9] using the SWE technique—this will be visited in proper detail under the section on analytical techniques—and soon followed by a review article [10]. Both works are concerned mainly with approximations to the Luneburg formula by the use of discretized structures, rather than the alternative approaches here called "nonuniform." Nevertheless, Schrank and Sanford [10] state that aperture efficiency in stratified lenses tends to be maximized by applying the criterion of equal projected area, rather than equal radial steps, of each layer, which

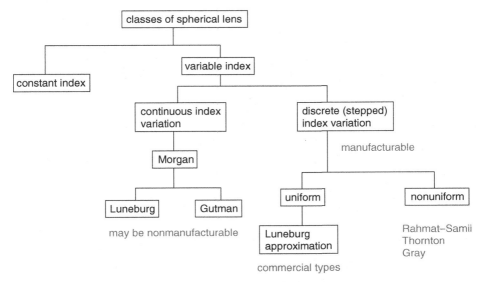

Figure 6.10. Spherical lens family tree.

indicates a move away from the very simplistic latter approach while retaining an adherence to the Luneburg curve of Equation (6.1). Notably, Sanford [10] also reiterates the advantages of quite small radial steps of as little as a quarter wavelength, and similarly, Reference 11, which reports a practical Luneburg lens array demonstrator, cites maximum steps of around one wavelength. These statements on shell thickness have been established practice until work on two- and three-layer lenses of the mid- and late 2000s.

6.3 ANALYTICAL METHODS

6.3.1 Ray Tracing

Ray-tracing models can offer some insight into the properties of spherical lenses as collimating apertures. The technique is quite tractable for constant-index spherical lenses but less so for stepped-index lenses. For the former case, the theta- and phi-directed electric fields are [12]

$$E_\theta = E_0 \frac{\cos\theta\sin\phi}{\sqrt{1-\sin^2\theta\sin^2\phi}} \sqrt{2(m+1)} \left(\cos\theta\right)^{\frac{m}{2}} \qquad (6.3)$$

$$E_\phi = E_0 \frac{\cos\phi}{\sqrt{1-\sin^2\theta\sin^2\phi}} \sqrt{2(m+1)} \left(\cos\theta\right)^{\frac{m}{2}} \qquad (6.4)$$

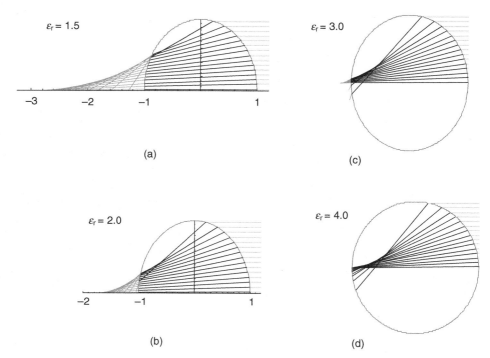

Figure 6.11. Ray tracing in dielectric spheres.

for a y-polarized primary feed whose pattern is

$$G(\theta) = 2(m+1)(\cos\theta)^{m} \tag{6.5}$$

over half of space ($\theta \le \pi/2$) and zero elsewhere.

Figure 6.11 shows the results from a ray-tracing program for four cases of spherical lens with different values of dielectric constant, increasing respectively from 1.5 to 4.0 in Figure 6.11a–d. In each case, a plane wave is incident on the right-hand side. In Figure 6.11a,b, the linear dimension is relative to the lens radius. From Reference 12 the paraxial focus, normalized to lens radius, is given by

$$f = \frac{\eta}{2(\eta - 1)}, \tag{6.6}$$

where refractive index $\eta = \sqrt{\varepsilon_{\mathrm{r}}}$.

In Figure 6.11a, where the index is lowest, paraxial rays arrive at the paraxial focus at 2.72 radii. Nonparaxial incident rays emerge in a region progressively closer to the lens outer edge, shown by the exit lines on the left side. Thus, the constant-index spherical lens does not exhibit a unique focus in the manner of a parabolic reflector or a

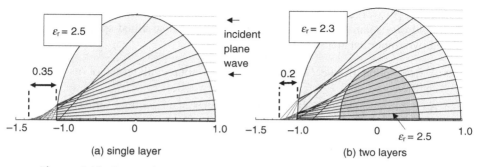

Figure 6.12. Ray tracing comparing (a) a single-layer and (b) a two-layer lens.

focusing lens. Rather, a quasi-focal region exists. As DC increases to 2 in Figure 6.11b, this region becomes more compact and, as a consequence, the lens exhibits a shorter f/D ratio and is also expected to exhibit higher aperture efficiency. Now increasing the constant to 3, in Figure 6.11c, the paraxial focus is quite near the lens edge and non-paraxial rays are focused at points inside the lens. For DC = 4, the paraxial focus lies at the lens edge.

Figure 6.12 compares the focal region of a constant-index lens with that of a stepped-index type, which is also discussed in a later chapter. In each case, a plane wave is again incident on the right-hand side. Here we see how the quasi-focal region is rendered more compact when a second layer is adopted and the dielectric constant values have been chosen to be favorable. This would lead us to expect a more efficient focusing power and thus higher aperture efficiency, although thus far, we have presented ray tracing only as an illustrative tool and have not yet developed a full electromagnetic analysis.

While ray tracing has been used as an approximate analytical tool for constant-index lenses (e.g., see Reference 12), it is unwieldy for stepped-index types, which we will be largely concerned with. As a consequence, the topic will not be covered in any more detail and instead we will proceed to modal analysis as discussed in the following section.

6.3.2 SWE

Also called modal expansion, since the spherical vector wave modes of free space form the basis functions of the method, the mathematics and notation was thoroughly described by Stratton [13] and from which scattering of plane waves by either metallic or dielectric spheres may be derived. This, in fact, was predated by Mie's analysis, famous for showing that the blueness of the sky is due to the strong dependence on the wavelength of scattering of the optical spectrum by small atmospheric particles. Stratton provides a clear notational framework for the description of the spherical vector wave modes, but his syntax is by no means encountered uniformly across other literature.

Sanford [9] worked within Stratton's notational framework to describe scattering of the field of a small elemental dipole incident upon a stratified spherical lens. This entailed two important steps:

- A description of the SWE of the dipole source in terms of its expansion coefficients
- An iterative derivation of the scattering matrix of an arbitrary series of concentric dielectric shells

The SWE method is well suited for implementation in computer code since the source and scattered fields are expressed naturally as a summation of angle-dependent terms. While all that is needed is, in essence, contained in Sanford's paper, albeit augmented by Stratton for clarification of notation and meanings of terms, some care should be exercised if attempting to use these works for this purpose. First, the syntax contains much minutiae of detail and, second [9], contains a small number of typographic errors. From Ludwig [14], who also prefaces a short treatise on spherical waves with a note of caution concerning nonstandardization of notation, the general form for the expansion of an electromagnetic field is given by

$$E = E_0 \sum_{n=1}^{\infty} \sum_{m=0}^{n} \left(a_{{}_{e_o}mn}^{(i)} \mathbf{m}_{{}_{e_o}mn}^{(i)} + b_{{}_{e_o}mn}^{(i)} \mathbf{n}_{{}_{e_o}mn}^{(i)} \right) e^{j\omega t} \tag{6.7}$$

for the electric field, and

$$H = H_0 \sum_{n=1}^{\infty} \sum_{m=0}^{n} \left(a_{{}_{e_o}mn}^{(i)} \mathbf{n}_{{}_{e_o}mn}^{(i)} + b_{{}_{e_o}mn}^{(i)} \mathbf{m}_{{}_{e_o}mn}^{(i)} \right) e^{j\omega t} \tag{6.8}$$

for the magnetic field. The engineering notation for time dependency, $e^{j\omega t}$, is shown in Equations (6.7) and (6.8) but, for brevity, may be dropped or subsumed into other constant terms, E_0 and H_0, without loss of generality.

Equations (6.7) and (6.8) can be interpreted as slightly analogous to Fourier series with which the general reader may be better acquainted: The modes $\mathbf{m}$ and $\mathbf{n}$ are the basis functions, and the coefficients a and b are the weights. Modes $\mathbf{m}$ and $\mathbf{n}$ are terms that are dependent on the angles θ and ϕ and distance r, as we shall see below, while the weighting coefficients are just complex numbers without any coordinate dependency.

Now, Equations (6.7) and (6.8) are in very general and condensed forms, and so some further commentary is warranted at this stage before presenting variants of the equations better suited to our analysis of spherical lenses.

First, care over syntax! The ${}^e_o$ syntax refers to even (e) and odd (o) modes, respectively, while the 0 occurring in E_0, H_0, and the summation terms is numeric zero.

$\mathbf{m}$ and $\mathbf{n}$ are then the spherical vector wave modes, which are functions of the modal numbers m and n, respectively, the azimuthal (ϕ) and polar (θ) wave orders. $\mathbf{m}$ and $\mathbf{n}$ are also encountered in various flavors, even and odd, and four other types denoted by (i), which represents the type of Bessel function z_n encountered in

$$\mathbf{m}_{\substack{e \\ o}mn} = \mp \frac{m}{\sin\theta} \overset{(i)}{z_n}(kr) P_n^m(\cos\theta) \frac{\sin}{\cos}(m\phi)\hat{\theta}$$
$$- \overset{(i)}{z_n}(kr)\frac{\partial}{\partial\theta} P_n^m(\cos\theta) \frac{\cos}{\sin}(m\phi)\hat{\phi} \tag{6.9}$$

$$\mathbf{n}_{\substack{e \\ o}mn} = \frac{n(n+1)}{kr} \overset{(i)}{z_n}(kr) P_n^m(\cos\theta) \frac{\cos}{\sin}(m\phi)\hat{r}$$
$$+ \frac{1}{kr}\frac{\partial}{\partial kr}\left(kr\,\overset{(i)}{z_n}(kr)\right)\frac{\partial}{\partial\theta} P_n^m(\cos\theta) \frac{\cos}{\sin}(m\phi)\hat{\theta} \tag{6.10}$$
$$\mp \frac{m}{kr\sin\theta}\frac{\partial}{\partial kr}\left(kr\,\overset{(i)}{z_n}(kr)\right) P_n^m(\cos\theta) \frac{\sin}{\cos}(m\phi)\hat{\phi},$$

where

$$k = \frac{2\pi}{\lambda_0}\sqrt{\varepsilon_r},$$

λ_0 is the free-space wavelength,

$P_n^m(\cos\theta)$ is the associated Legendre polynomial,

and $\hat{r}$, $\hat{\theta}$, and $\hat{\phi}$, respectively, are radial and angular unit vectors in spherical coordinates. Derivates of terms are also encountered. Note the dependency of k on dielectric constant ε_r for the material in which the equations hold.

Equations (6.7) and (6.8) are now very close to Sanford's notation, although we are here using lowercase z, and remain mindful of the several definitions of $\overset{(i)}{z_n}(kr)$:

$\overset{(1)}{z_n}(kr) = j_n(kr)$, the spherical Bessel function of the first kind (6.11)

$\overset{(2)}{z_n}(kr) = y_n(kr)$, the spherical Bessel function of the second kind (6.12)

$\overset{(3)}{z_n}(kr) = \overset{(1)}{h_n}(kr)$, the spherical Hankel function of the first kind (6.13)

$\overset{(4)}{z_n}(kr) = \overset{(2)}{h_n}(kr)$, the spherical Hankel function of the second kind. (6.14)

The regions of validity for Equations (6.11)–(6.14) are described by Ludwig [14], who states that, in general, two of the four functions are required to completely describe a radiation problem, which we will find to be the case for scattering of a source field by a spherical lens. For this problem, the Bessel functions apply in a sphere whose radius is less than or equal to the source radius, while the Hankel functions apply for the region of space exterior to this, a situation loosely illustrated in Figure 6.13.

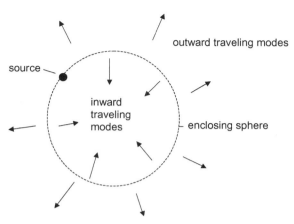

Figure 6.13. Pictorial representation of regions of validity for spherical vector wave modes, after Ludwig. (The arrows are not trying to represent rays.)

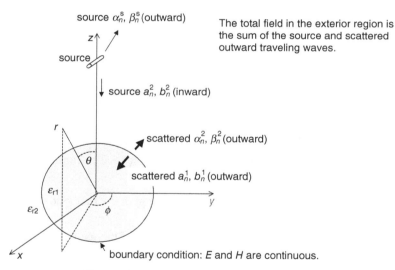

Figure 6.14. Scattering in a dielectric sphere.

Now applying the SWE formulation to scattering from a dielectric sphere, and starting with a single-layer (constant-index) case, the source and scattered terms are illustrated in Figure 6.14 and where the source is a short x-directed dipole. Owing to the rotational symmetry of the source, we have $m = 1$ and only a single summation in n is required. This convenient simplification applies to other source types with spherically symmetrical patterns and arises from mode orthogonality properties [15]. Bearing in mind a superposition of both inward traveling (1) and outward traveling (3) mode types, Equations (6.7) and (6.8) then reduce to

$$E = E_0 \sum_{n=1}^{\infty} \left(a_n^P \mathbf{m}_{o1n}^{(1)} + b_n^P \mathbf{n}_{e1n}^{(1)} + \alpha_n^P \mathbf{m}_{o1n}^{(3)} + \beta_n^P \mathbf{n}_{e1n}^{(3)} \right) \tag{6.15}$$

for the electric field and

$$H = H_0 \sum_{n=1}^{\infty} \left(a_n^P \mathbf{n}_{o1n}^{(1)} + b_n^P \mathbf{m}_{e1n}^{(1)} + \alpha_n^P \mathbf{n}_{o1n}^{(3)} + \beta_n^P \mathbf{m}_{e1n}^{(3)} \right) \tag{6.16}$$

for the magnetic field. These closely resemble Sanford's notation.*

Here P denotes a spherical "shell" region, since the expansion coefficients for source terms a, b and scattered terms α, β are specific to a given region in space. As such, P is just a counter, with $P = 1$ for region 1, $P = 2$ for the next concentric region, and so on. Figure 6.14 shows the two regions of this simple case—region 1 is inside the dielectric sphere and region 2 is in free space.

In general, any number of regions may be accounted for and each set of coefficients derived iteratively from the boundary conditions following the recipe of Sanford [9]:

$$a_n^{P-1} = U_n^P \cdot a_n^P$$
$$\alpha_n^{P-1} = U_n^P \cdot \alpha_n^P, \tag{6.17}$$

where U_n^P is a term derived from the E-field boundary condition at the boundary between layers P and $P - 1$. A very similar and concise formulation uses the H-field boundary condition to derive a scattering matrix, V_n^P, which relates terms b and β. To derive these matrices, the reader may use Sanford's compact formulae [9] or may be inclined to derive them from first principles. The latter route offers an insightful tutorial exercise, and a sanity check for the simplistic single-layer sphere may be found in chapter IX of Stratton [13], but a full transcript of the method is rather beyond scope here.

It is worth adding that the source and scattered fields loosely illustrated by arrows in Figure 6.14 should not be interpreted as rays: The inward traveling modes in the central region may be thought of as being absorbed at the origin, and there are never outward traveling modes in the central region since there is no source or boundary inside this region.

A similar treatment of the derivation of the scattering matrices may also be found in Reference 16, which also describes how outer layers (e.g., an antenna radome) may be included, along with some comments on processor time for computing radiation patterns for electrically large lenses. Thornton [16] also presents some measurements for a lens antenna with a protective environmental cover. Such a cover layer would be

---

* Sanford [9] erroneously has the equivalent of $a_n^P \mathbf{n}_{e1n}^{(1)} + b_n^P \mathbf{m}_{o1n}^{(1)}$ as the first two terms in the expansion for H, the e and o having been transposed in a typographic error. Another error is encountered in the ϕ dependent term of $\mathbf{n}_{emn}$, the sin and cos terms having been switched.

needed for many applications, for example, when the antenna is mounted on a moving vehicle.

6.3.2.1   *Far Field.* The above wave expansion formulation is a very powerful analysis technique for stepped-index spherical lenses. Setting aside for now the simplistic feed model (an improved model will be discussed later), its strengths are that the electric and magnetic fields can be evaluated anywhere, that is, at any arbitrary angle and distance in free space, or indeed within any of an arbitrary number of different dielectric layers, P. As such, Equations (6.15) and (6.16) are still expressed in fairly general terms so that they maintain this wide-reaching applicability.

Oftentimes, however, the engineer will be more interested in only the far-field patterns, so that directivity and sidelobe levels may readily be evaluated. In this case, little more may be needed than a version of Equation (6.15), where we evaluate only the sum of source and scattered outward traveling modes in the outermost or Nth region:

$$E^N = E_0 \sum_{n=1}^{\infty} \left(\alpha_n^S + \alpha_n^N \right) \mathbf{m}_{o1n}^{(3)} + \left(\beta_n^S + \beta_n^N \right) \mathbf{n}_{e1n}^{(3)}. \tag{6.18}$$

Here, the superscript S refers to the source coefficients (see Fig. 6.14). For far-field calculations, the asymptotic solutions for Equations (6.11)–(6.14) may be used, where r dependency vanishes. Furthermore, it is common practice to examine single planes, or cuts, of the spatial far-field pattern: These are most often the so-called principal planes, the E-plane ($\phi = 0°$) and H-plane ($\phi = 90°$), although any plane may be chosen. This leads to a further simplification because ϕ-dependent terms reduce to constants in the mode expressions in Equations (6.9) and (6.10).

6.3.2.2   *Convergence.* The wave expansion (summation of terms) must, in practice, be truncated at some finite value. Ludwig [14] proposes that the expansion can be truncated, as a rule of thumb, at

$$N_{\text{sum}} = kr_o, \tag{6.19}$$

where r_o is the radius of a sphere enclosing the source. From Figure 6.14, we see that this will be at least the spherical lens radius or, more likely slightly larger than this. Thus (and obviously), larger structures require a longer summation of terms, impacting directly on processor time when the above formulae are implemented in computer code. Some results using such code are presented in the following section and where convergence is explored by looking into the rate of decay of the expansion coefficients' magnitude.

6.3.3   Computational Method and Results

Code has been developed by the author over a period of several years to implement the above SWE formulations using the package *Mathematica*. A suite of programs derive far-field radiation patterns of arbitrary, multilayer spherical lenses based on a small number of input parameters.

These input parameters can be entered in list form:

- the shell radii, followed by the radius of the first feed dipole,

$$r^p = \{r_1, r_2, r_3, \ldots r_{feed}\},$$

and

- the dielectric constant (DC, now adopting a shorthand notation) of each layer P. The final entry in the list is always unity for free space:

$$DC^p = \{DC1, DC2, DC3 \ldots 1\},$$

so that the final layer is air.

A single dipole feed is rather too simplistic for these purposes because it is not representative of a directional primary feed. An arbitrary primary feed may, in principle, be synthesized from a large number of elementary dipoles, but the associated computational complexity and processor time can become excessive. A pragmatic primary feed model can use just two dipoles as illustrated in Figure 6.15 to yield an end-fire array whose main lobe is directed toward the spherical lens along the z-axis. This feed pattern exhibits a null in the opposite direction $(-z)$ and so is reasonably representative of a wide-angle feed illuminating the sphere. The technique does not accurately model real primary feeds but does allow for a valid comparison of different lens geometries while retaining computational tractability. For clarity, Figure 6.15 shows only a single dielectric layer in free space.

Using the above input data, a software routine derives the coefficient sets a, b, α, β for the outer region, each list having N_{sum} entries as the truncation limit for the

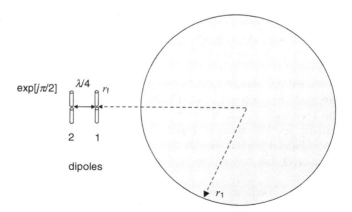

Figure 6.15. Feed model with constant-index lens.

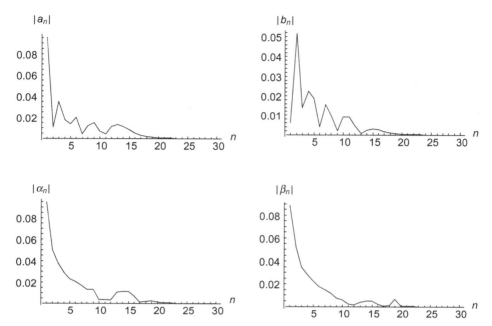

Figure 6.16. Form of expansion coefficients for small constant-index lens of two-wavelength radius, two-dipole feed model.

summation. By way of example, the form of these coefficients is explored in Figure 6.16, where the structure analyzed is a constant-index lens with DC = 2.3, $r_1 = 2$, and $r_{\text{feed}} = 2.5$. (The radial dimensions are expressed in wavelength.) Figure 6.16 shows the absolute values of the complex expansion coefficients for the set of coefficients associated with the first feed dipole only.

The form of the results in Figure 6.16 may not be of immense interest save for the following: The magnitude of the coefficients can oscillate quite strongly, and the terms have decayed to near zero, in this particular example, by $n = 23$. From Equation (6.19), we would derive $N = 16$ for the series truncation if using the product $k \cdot r$ at the feed. In contrast, the product at the lens edge would yield $N = 19$ due to the higher value of dielectric constant, still an underestimate of the required number of terms.

Now moving onto an electrically larger problem, where the lens radius is 10 wavelengths and the feed radius 11.5 wavelengths, Figure 6.17 shows that terms have decayed to zero by $n = 80$, whereas Equation (6.19) would have 95. In practice, Equation (6.19) is indeed a good rule of thumb for large spheres (larger than a few wavelengths' diameter), but it is worth monitoring the decay of terms as a means of checking convergence.

Having derived the coefficient sets, the far-field (or near-field, if needed) patterns can be evaluated. While three-dimensional derivations are again time-consuming, much can usually be inferred from the principal planes. Figure 6.18 shows the E-plane

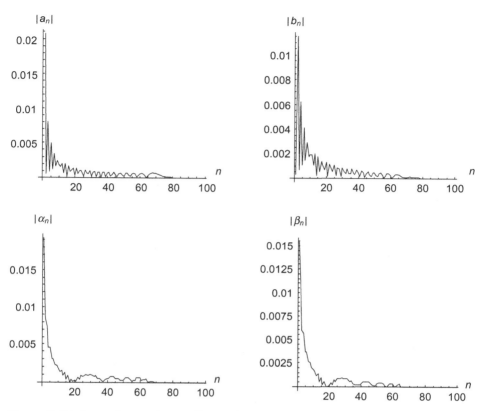

Figure 6.17. Form of expansion coefficients for constant-index lens of 10-wavelength radius.

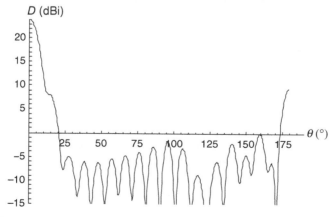

Figure 6.18. E-plane radiation pattern of two-wavelength radius lens derived from SWE.

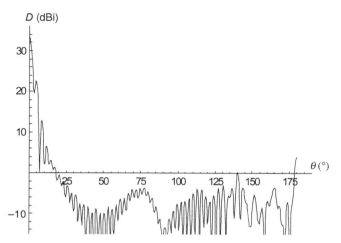

Figure 6.19. *E*-plane radiation pattern of 10-wavelength radius lens derived from SWE.

far-field radiation pattern of the two-wavelength radius lens, and Figure 6.19 that for 10 wavelengths. The scaling for directivity D in each case has been estimated by assuming rotational symmetry around azimuth ϕ—an approximation to reduce processor time. Accuracy for directivity scaling can be improved by using a more cumbersome derivation of the full three-dimensional pattern, or with little loss in accuracy but a significant reduction in processor time by interpolation between principal planes.

On a similar note, resistive losses can be accounted for in the SWE analysis. Thus far, the dielectric constant (DC or ε_r), for purposes of convenience and shorthand, has been expressed as a real number, this being entirely valid for lossless materials. In practice, loss tangent may also be included in the SWE method, this being readily accommodated in the SWE computer code reported.

Of further interest to readers may be examples of SWE terms found in Wood's work on reflector antennas [17].

6.3.4 Generic Feed Pattern

The above end-fire array feed model implemented in SWE clearly has limitations in terms of accurately accounting for real feed patterns. Other approaches to feed models have been reported in literature. In Sanford's 1994 work [9] on scattering in stratified lenses, the radiation pattern of a pyramidal horn was approximated as the sum of many short dipoles. A similar approach was reported in References 18 and 19 but with an emphasis on patch antennas and where the current on the patch antenna aperture was described as a set of short dipole currents. This entailed the extensive use of coordinate transforms to account for the displacement of each short dipole from the z-axis. (The two-element array pattern is more tractable and simplistic because each dipole lies on the z-axis and no further transform is required.) Mosellaei and Rahmat-Samii [6] used a method of intermediate complexity where four dipoles modeled the feed pattern.

It is of course entirely feasible to abandon the dipole as the building block of the feed pattern in the SWE scattering problem and to derive an expansion of any generic antenna whose radiation pattern is known. This method presents great flexibility since the feed pattern may be derived from measurement data. It is important that both phase and amplitude data are captured. Also, the measurement may take place in the near field and still be sufficient to fully derive the necessary expansion terms. Alternatively, a commercial electromagnetic solver package, or any numerical technique based along similar principles (finite difference time domain, transmission line matrix, and so on), may be used to model the feed pattern. The reader may well ask: Could not such a method be used to model the feed–lens combination in its entirety? The answer would be yes, in principle, but the electrical size of the problem and hence the processor and memory demands could, for a practical high-gain antenna, be vastly greater than the analysis of the feed on its own. This is discussed in a little more detail in the following section.

Returning to the SWE method applied to any general feed, the theory is presented by Ludwig [15],* where the expansion terms are

$$a_{e,omn} = \left(\frac{1}{z_n(kr)}\right)^2 \frac{2n+1}{\pi 2n(n+1)} \frac{(n-m)!}{(n+m)!} \int_0^{2\pi} \int_0^{\pi} -\mathbf{m}_{e,omn} \cdot E_t(r,\theta,\phi)\sin\theta \, d\theta \, d\phi \qquad (6.20)$$

$$b_{e,omn} = \left(\frac{1}{\frac{1}{kr}\frac{d}{dr}(rz_n(kr))}\right)^2 \frac{2n+1}{\pi 2n(n+1)} \frac{(n-m)!}{(n+m)!} \int_0^{2\pi} \int_0^{\pi} -\mathbf{n}_{e,omn} \cdot E_t(r,\theta,\phi)\sin\theta \, d\theta \, d\phi,$$

$$(6.21)$$

where $E_t(r,\theta,\phi)$ is the tangential electric field of the source, which we wish to describe as an expansion. Radius r may be any near-field radius, in which case this value is used in the spherical Bessel function $z_n(kr)$, or may be taken as infinity in which case the far-field version for $z_n(kr)$ may be used.

Thus, if $E_t(r,\theta,\phi)$ is known, its expansion terms may be derived. The reader is referred to Reference 15 for a more detailed discussion. Also Reference 20 may be helpful since it presents a very similar analysis for the approximate expansion of a Gaussian beam.

Nevertheless, Equations (6.20) and (6.21) are not trivial to implement in computer code. A "brute force" approach is to implement the integration in a numerical package. A similar expansion method is used in Reference 21 but working with source equivalent currents, and later applying the source field to either spherical stratified (Luneburg) or hemispherically stratified (Maxwell fish-eye) lenses.

---

* Again, care must be taken with notation. Ludwig's version of the mode $\mathbf{n}_{e,omn}$ appears to have transposed cos and sin terms and the sign of the phi (ϕ) vector component.

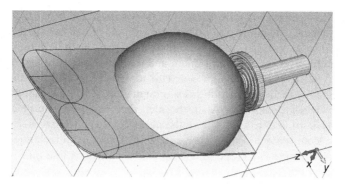

Figure 6.20. Feed, hemisphere lens, and ground plane, from CST MWS 3-D model.

6.3.5 Commercial Solvers

Electrically large antenna problems are among the most challenging for commercial solvers such as CST MICROWAVE STUDIO (CST MWS) [22] or Ansoft HFSS. Many other products are available. Gray et al. [23, 24] reported results for feed-(spherical) lens interaction problems using FEKO™ for lens diameters up to about eight wavelengths. Typically, as diameter increases, the number of solver mesh cells increases roughly in a cubic relationship from a consideration of the volume of space around the antenna.

One example of a lens antenna model found tractable using CST MWS (2009 version) is illustrated in Figure 6.20. This considers a hemisphere with ground plane, a topic discussed in more detail in Chapter 7. The ground plane dimension was 8×12 wavelengths. Using the time-domain solver and a slightly coarse mesh of 10 lines per wavelength, the total number of mesh cells was around 4 million. A basic laptop computer with 2-GHz processor and 2-GB RAM memory took about 60 minutes to arrive at a convergence of −20 dB in the time-domain calculation. This represents a fairly low level of accuracy. However, better computers (>24-MB memory would be recommended) would yield much reduced solve times or a better accuracy for a given elapsed time.

A snapshot of some of the output data from CST MWS is shown in Figures 6.21 and 6.22. Owing to the low level of convergence in this example, the plot of S11 shows little structure, although the mean value in the region −17 dB is realistic.

A short treatise on commercial solver run times is also found in Reference 21, where CST MWS and HFSS are compared. Maximum lens diameters were limited to 16 and 10 wavelengths, respectively. The run times were presented as a comparison of the commercial solver compared to bespoke mode matching code: The latter always outperformed the former by a factor of at least 2 but increasing to infinity for the larger lens dimensions where the commercial solvers failed to converge.

A powerful feature of solvers is the ability to make small, incremental changes to structures to examine trends in simulated antenna performance. Figure 6.22 shows

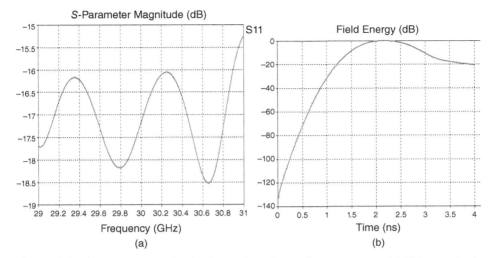

Figure 6.21. CST derived results for hemisphere lens-reflector antenna (a) S11 magnitude and (b) time-domain energy partial convergence.

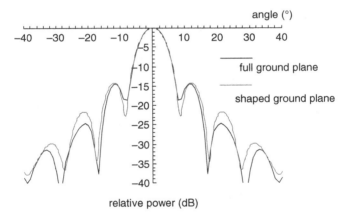

Figure 6.22. CST MWS derived radiation patterns of hemisphere lens for two ground plane geometries.

how shaping the ground plane from square corners to radiused corners introduces a small degradation in the radiation pattern. This was manifested as about a 0.1-dB reduction in gain and a very slight growth in sidelobes. This small effect is expected since the ground plane and feed orientation used in Figure 6.20 has been chosen to be close to the minimum ground plane length (derived from basic geometric optics considerations) that is needed to fully recover the image of the hemisphere (see Chapter 7). Thus, any reduction in ground plane area is expected to give rise to a degradation in gain.

A promising area of research in the field of spherical lens antennas utilizes a combination of commercial solvers with SWE techniques. The strength of the former, in this context, would be accurate analysis of feed radiation patterns. The strength of the latter is fast analysis of lenses with spherical symmetry. Ground plane effects can be treated by hybrid techniques including near–far transforms, a topic taken up in the following chapter. Some work using solvers to extract feed patterns has already been reported: Fuchs et al. [21] allude to "commercial solvers," while Nikolic and Weily [25] used CST MWS [22] to simulate the feed pattern. In Reference 25, the expansion coefficients for the feed pattern were derived from an integration over a planar surface in front of the radiating feed aperture, which contrast with Ludwig's integrals (Eqs. 6.20 and 6.21), which are over an enclosing sphere. The planar approximation is of course an entirely valid one for a directive antenna where the vast majority of the radiated energy is incident upon the planar aperture chosen, mirroring quite nicely the planar near-field measurement technique encountered briefly in the closing section of Chapter 1.

6.4 SPHERICAL LENS MATERIALS AND FABRICATION METHODS

Low-volume lens manufacture can be quite challenging. Polymers have been used very successfully, but these present a limited range of options for dielectric properties. For lens diameters greater than about 0.5 m, the mass of a polymer lens becomes significant or, at least, too large to be easily handled by one person. For stepped-index or multilayer spherical lenses, the outer layers (being the largest) are significantly more difficult to fabricate than any inner layers. Materials and fabrication techniques are closely related. A review of lens manufacturing techniques was carried out during the development of a lens-reflector antenna for satellite communications, discussed in Chapter 7, prior to which a two-layer stepped-index polymer lens of about 240-mm diameter had been constructed.

Various approaches to spherical lens manufacture fall into two main categories, namely,

- mechanical machining from material billet and
- molding.

6.4.1 Machined Polymers

Conceptually straightforward, this method also leads to the first successful stepped-index two-layer hemisphere lens described in Chapter 7. Machine tools applicable to the task include lathes and three-axis mills, which might in either case be under computer numerical control (CNC). Before machining, a polymer billet must first be secured, and this might entail attachment to a backing plate. Slightly different approaches apply to inner or outer layers in multilayer lenses. A sequence of steps toward machining a polymer hemisphere is illustrated in Figure 6.23. Here, a jig serves as a mechanical interface between a lathe chuck and the billet. Holes are drilled and tapped in the

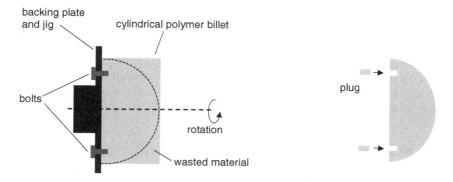

Step 1. Mount and machine billet   Step 2. Plug holes

Figure 6.23. Machining steps for hemisphere lens.

Figure 6.24. Machining a hemisphere lens from a Rexolite billet.

billet so that it can be attached to the backing plate of the jig. The jig and billet are then machined to produce the hemispherical surface. Later, the holes in the flat face of the hemisphere need to be plugged. Since the holes are tapped, it can be useful to drill out the tapped walls of these holes with a flat-bottomed cutter. Then, cylindrical plugs can be inserted into the cylindrical holes, eliminating as far as practically possible any air gaps.

A Rexolite billet is shown in Figure 6.24, here attached to a backing plate and being machined on a lathe. The lathe cutter follows a profiled tool produced on another CNC machine. A coolant is introduced to the surface being machined. The hemisphere shown in Figure 6.24 has a diameter of about 300 mm.

On increasing the dimensions, a difficulty with a machined part of this type is procuring a sufficiently large billet of solid polymer. Such a billet is also costly and material tends to be wasted. An alternative approach to machining a single piece would be fabrication of smaller sections of the outer layer, either as vertical slices or quadrants/ octants, and so on. A disadvantage of this approach is the need to bond together the pieces and the likelihood of introducing air gaps, which tends to negate the advantage of using as few material layers as possible.

6.4.2 Molding

There are several approaches to molding, depending on the type of material chosen. While low-loss polymers are used in industrial injection molding processes for relatively thin-wall components, it is doubtful that this would scale successfully to the relatively large, solid items such as hemispheres. In any case, fabrication of the mold for this process could be prohibitively expensive, at least for prototyping purposes. Injection molding is generally recognized as a cost-effective means of mass-producing relatively low-cost consumer items, but one where setup and tooling costs dominate.

Syntactic foam is a promising material for lens antenna application. It consists of microscopic spherical air-filled glass beads bound in a type of epoxy material. A proclaimed advantage of this material is that the dielectric constant can be controlled by adjusting the relative ratio of glass beads to epoxy as well as the glass bead dimension. The epoxy may be cured in a mold to produce a finished part or a billet, which may later be machined.

6.4.3 Polymer Foams

Common polymer foams include expanded polystyrene and polyethylene (PE), where the finished material is a foam/air mixture. Typically, materials are very lightweight and dominated by air and are therefore of very low dielectric constant ε_r (slightly greater than 1). In theory, however, a blend for ε_r between 1 and the ε_r value for the polymer can be obtained. The simple mixing rule is

$$\varepsilon_{rf} = 1 + (\varepsilon_{rp} - 1)\frac{\rho_f}{\rho_p}, \qquad (6.22)$$

where

ε_{rf} is the dielectric constant of the foam material,
ε_{rp} is the dielectric constant of the raw polymer,
ρ_f is the density of the foam, and
ρ_p is the density of the raw polymer.

Where a low-density foam is required, it is clearly advantageous to use a polymer raw material with a low ratio of density to dielectric constant so as to minimize the lens

TABLE 6.1. Properties of Common Polymers

	ε_{rp}	ρ_p
High-density polyethylene (HDPE)	2.3	0.95
Polystyrene (PS)	2.5	1.05
Polypropylene (PP)	2.3	0.9
Polyurethane (PU)	4.0	Ether base[a]: 1.1–1.18
		Ester base: 1.2–1.26

[a] *Sources*: SD Plastics, San Diego, USA, http://www.sdplastics.com; also Dow Chemicals.

TABLE 6.2. Summary of Foam Densities and Dielectric Constants

Nominal Density (kg/m^3)	Nominal ε_r	Measured Density (kg/m^3)	ε_r Derived from Measured Density
300	1.76	320	1.81
420	2.07	410	2.04
520	2.32	499	2.27
650	2.65	619	2.57

mass. A few candidate materials encountered in commercial foam manufacturing processes are noted in Table 6.1.

Of these, polyurethane (PU) would offer the most favorable density properties. PU foam is also very common in industrial products. Unlike the injection molding process, where high temperatures and pressures are encountered, PU foam is made in much lower-cost processes and molds can be quite thin walled.

To further explore the dielectric properties of PU, samples were obtained from a supplier. From the above and assuming $\rho_p = 1.18$ leads to the theoretical properties shown in Table 6.2. These were labeled in terms of a notional density, that is, the supplier's value, which was later checked by measurement.

These densities were specified so as to capture the range of ε_r values that can be obtained using nonfoam polymers and also to establish a correlation between theoretical ε_r and measured values. Dielectric constant ε_{rf} was measured using a resonator technique, albeit at lower microwave frequencies than those that were later required. An improvised method [26] used a microstrip resonator in direct contact with a slab of the dielectric material and thus obviated the need to construct an enclosed resonator such as a waveguide type. Extensive modeling with a solver CAD package allowed the thickness of the material above the planar circuit to be accounted for. From simulation, it was found that beyond a few millimeters or so of thickness, the resonant frequency became insensitive to further increases. Thus, shift in resonant frequency versus ε_r could be derived with reasonable confidence.

The results [26] are summarized in Table 6.3. Known materials Rexolite and PE were also measured so as to provide calibration standards. The purpose of this

TABLE 6.3. Measured ε_r Values for PU Foams

Sample	ε_r Derived from Measured Density	Measured ε_r	
		5.8-GHz Resonator	7.8-GHz Resonator
PU 300	1.81	1.88	1.80
PU 420	2.04	1.99	1.96
PU 520	2.27	2.20	2.21
PU 650	2.57	2.65	2.57
	Nominal ε_r		
Rexolite	2.53	2.56	2.50
PE	2.3	2.29	2.32

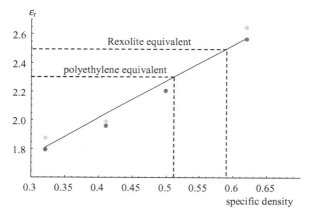

Figure 6.25. Measured ε_r versus specific density, polyurethane foam (solid line, theoretical curve).

investigation was then to synthesize equivalent materials to these known solid polymers using PU foam.

In Figure 6.25, the linear least squares fit is shown against measurement data for ε_r versus density. The target ε_r values for Rexolite and PE material equivalents have been captured.

6.4.4   PU Dielectric Loss

The resonator measurements indicated a detrimental property of the flexible PU foam samples, that is, dielectric loss. Again, this can be estimated with reasonable accuracy by comparing the measured spreading of the resonance response in the frequency domain (Q-value reduction) against theoretical response from simulation of the resonator circuit for various values of loss tangent. This approach leads to an estimate for loss tangent of 0.015 for the highest-density sample. In comparison, the loss tangent for PE and for Rexolite (while this is not known very exactly) is of the order less than 0.0002, and the measurement technique was thought too coarse to shed any further light on this figure.* To put these figures in context, the theory for spherical lens radiation performance shows that loss tangents of the order 0.0002 or less have an insignificant effect and the absolute accuracy of the parameter is not very relevant. However, for loss tangents greater than about 0.005, radiation performance starts to be significantly affected owing to the combined phenomena of absorptive loss and directivity reduction due to defocusing.

These findings suggest that PU foam obtained would be an unsuitable candidate for dielectric lenses in general. Nonetheless, it is possible that polymer foam materials would be worthy of future study and a ripe area for collaboration between microwave and materials engineers.

6.4.5 Artificial Dielectrics

The problem of dielectric loss in PU is also a conclusion echoed by Donelson et al. [27] and Kot et al. [28] from work by researchers at Australia's Commonwealth Scientific and Industrial Research Organisation (CSIRO) where spherical lenses were being considered as possible antenna elements for the square kilometer array (SKA) radio telescope. They proposed that aperture diameters up to about 1 m represented a practical upper limit for polymer lenses. More advanced materials, which might be called "artificial dielectrics," were also put forward in this context. Here [28], the goal was a reduced mass dielectric material with a controllable constant and suitable for lower cost and larger volume production, and larger diameter lenses than is practicable for more conventional polymers. The type of material developed, at least partially, was ceramic-loaded low-density foam. The ceramic used in practical experiments was rutile (TiO_2). This was mixed in a 1% volume ratio with very low-density polypropylene foam. Concentric hemispherical shells were not produced directly in a molding process but rather they were assembled from many sections or "tiles." Problems encountered included a difficulty in controlling the uniformity of the combined rutile/foam material and a sensitivity to the aspect ratio and orientation of the rutile particles, which could be realized as disklike or cigar-like in shape. Near-field radio frequency measurements revealed anisotropy in the one experimental lens that was produced, leading to mediocre efficiency. Unfortunately, this very promising line of inquiry appears to have not been taken any further. However, it would seem that, in principle, this approach to artificial dielectric materials could offer great dividends if it could be developed to sufficient maturity.

6.5 REVISITING THE CONSTANT-INDEX LENS

The SWE method described above has been a very useful and powerful computational tool in analyzing spherical lenses. Before going on to discuss multilayer lenses, it is

---

* An even more pragmatic "measurement" for dielectric loss is to place the samples in a microwave oven, preferably with a glass of water. While the water and PU warm up nicely, our "calibration" materials (PE and Rexolite) remain reassuringly cool.

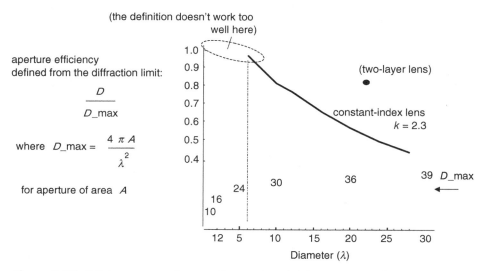

Figure 6.26. Efficiency versus diameter (in wavelengths) for constant-index spherical lens.

worth reflecting on the properties of constant-index spherical lenses [29], particularly noting that the computed directivities of the above two examples (Figs. 6.18 and 6.19) are rather high considering that this class of lens is often perceived as not particularly efficient. We have seen from ray-tracing drawings that the constant-index lens does not exhibit a unique focal point, and so only mediocre efficiency may be expected. However, the SWE results give more accuracy and a proper insight into the properties of this class of antenna. For the 10-wavelength radius example above, the computed directivity is 33.3 dBi, albeit including some approximations and a not wholly realistic feed model. This represents an area efficiency of 54%, while for smaller lenses, the efficiency is higher still. Figure 6.26 shows computed aperture efficiency, as a function of lens diameter, for the constant-index lens model considered above. This was first reported in Reference 30, and the graph was used in a similar form also in Reference 31.

What is notable here is that a quite respectable efficiency can be produced from a constant-index lens, up to, say, 35 dBi. For directivity up to about 27 dBi, we might expect a negligible benefit in using a stepped-index lens and where the cost would in all likelihood not be justified. For 30–40 dBi, an increasing benefit is seen by using a stepped-index lens in place of a constant-index lens since the efficiency of the latter falls away quite sharply in this regime. However, here a two layer lens can work very well. It might be inferred that the complexity of a multilayer (e.g., eight or more) lens according to received wisdom may be quite unnecessary even up to about a 60-wavelength diameter, at least if area efficiency is the main design objective. Of course, depending on the physical size of the lens, weight may well be a more critical factor and, as we have seen, the Luneburg lens (or the stepped-index approximation to it) will tend to use the lowest values of dielectric constant and so also minimize mass.

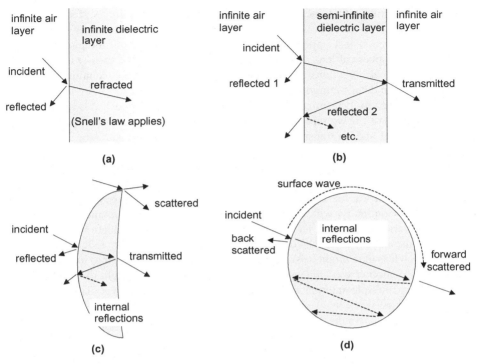

Figure 6.27. Wave scattering at dielectric surfaces.

However, for millimeter wave applications where antenna apertures may not be physically large, polymer lenses of 1, 2, or 3 layers can offer very practical solutions.

Another question often encountered relates to reflection losses: A perceived disadvantage of polymer lenses compared to traditional foam Luneburg lenses is that the surface reflection from the former must be higher because the dielectric constant is higher. However, this need not necessarily be the case, as the arguments and illustrations of Figure 6.27 will explain.

In Figure 6.27a, the situation is entirely straightforward: The magnitudes of the scattered fields or, to use transmission line terminology, their reflection coefficients, are each simple functions of the change in impedance at the boundary. There is no edge scattering to consider, nor multiple internal reflections. In Figure 6.27b, there are two planar boundaries, and the reflection coefficient is now also a function of the width of the dielectric slab: Should the width equal one-half of the wavelength in the dielectric medium, the reflections at each surface cancel due to their phase difference. This, of course, is the principle of the half-wave radome and resembles also the quarter-wave impedance transformer encountered in transmission lines, or of antireflection coatings encountered in optics. Now moving on to the lens in Figure 6.27c, the dielectric region is now finite and so exhibits edge scattering as well as the multiple reflections at both surfaces. A ray-tracing approach is useful in determining the required surface curvature according to Snell's law, but we should not expect it to give an accurate insight into the overall reflection loss experienced by incident rays. Distorting the picture further to arrive at the spherical lens shown in Figure 6.27d, we see that there will be a multiplicity of internal reflections and scattering in all directions. Ray techniques here are not much help in determining reflection loss, and it is more fruitful to invoke the SWE method to accurately determine scattering in any direction. To push this analogy to a more extreme level, if we replace the dielectric in Figure 6.27a with a conductor, the reflection is total and no transmitted ray can exist on the right-hand side. In contrast, if we replace the dielectric sphere in Figure 6.27d with a conducting sphere, forward scattering can still occur, albeit perhaps at a low level depending on the electrical diameter of the sphere. The point here is that it is overly simplistic to imagine that the back-scattering in the lenses in Figure 6.27c,d is dominated entirely by a notion of impedance discontinuity, which is really more applicable to the geometries in Figure 6.27a,b. Nevertheless, higher DC values do in general increase reflection loss, particularly for constant-index lenses, but for lower constant materials, the reflection loss can be very much less than might be supposed from a simplistic (ray tracing) mental picture. Or, expressed another way, the spherical lens might be better thought of as a resonator that supports multiple internal and surface modes rather than as a slab of material reflecting a plane wave incident from an air layer in free space.

Figure 6.28 shows the effect of changing dielectric constant in a four-wavelength radius lens fed by a pair of short dipoles in accordance with Figure 6.15. (For the cases $\varepsilon_r = 2.0$ and 3.0, the gap to the first dipole is 0.5 λ; for the case $\varepsilon_r = 5.0$, the feed gap is reduced to 0.02 λ.) In Figure 6.28, we see some interesting trends: The efficiency of

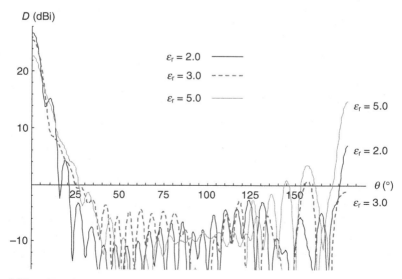

Figure 6.28. Effect of changing dielectric constant in a four-wavelength radius spherical lens.

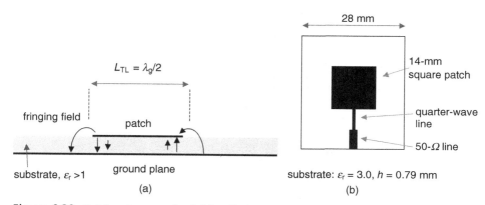

Figure 6.29. Patch antenna as feed: (a) radiation mechanism and (b) dimensions for 5.8-GHz center frequency.

the lens as a collimating aperture reduces as dielectric constant increases, but the back-scattered lobe (at $\theta = \pi$) does not at first increase as might have been expected from a consideration of the impedance mismatch.

6.5.1 A Practical, Patch-Fed Hemispherical Constant-Index Lens

For lower microwave frequencies (say, below 10 GHz), a patch antenna has been shown to be an effective feed for a spherical or hemispherical lens. In Reference 30, a practical arrangement was reported where a theme is the use of unlicensed radio bands below 10 GHz and communications to either aerial platforms or satellites. Here, a simplistic scanning antenna was proposed, which might use switched beams rather than a continuous mechanical scan, so as to cover a particular region of space with contiguous beams. Such an antenna could offer a quite low-cost solution, and furthermore, it was proposed that a single lens could be shared by primary feeds of various types producing beams at different frequencies in different regions of space.

A selection of printed antennas for operation in the region 5.0–6.0 GHz was fabricated. A simple, low-cost design for 5.8 GHz is shown in Figure 6.29b, and the fabricated item is seen on the left-hand side of Figure 6.30. Face down on the surface of the PE lens is another patch feed, this being used to illuminate the lens. The four-element patch array seen in the photograph was not used to illuminate the lens in this particular experiment but pictured to indicate the likely circuit dimensions and interelement spacing should a multifeed geometry be used. The lens diameter was 160 mm.

The measured E-plane and H-plane radiation patterns for the patch-fed lens of Figure 6.30 are shown in Figure 6.31. The overall gain was in the region of 17 dBi.

6.5.2 Off-Axis Array-Fed Spherical Lens

In Reference 32, the ray-tracing method of Schoenlinner et al. [12], for a Teflon constant-index spherical lens, was extended to allow for an offset d of the array feed

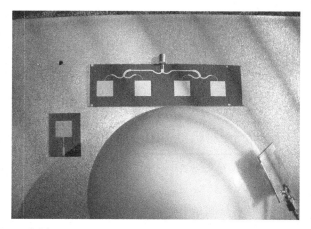

Figure 6.30. Components of patch-fed hemisphere lens antenna.

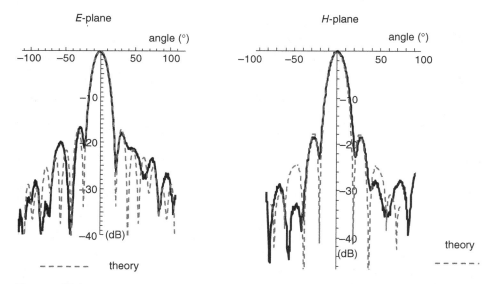

Figure 6.31. Measured and (dashed) theoretical radiation patterns in constant-index hemispherical lens.

axis with respect to the lens axis (Fig. 6.32). This allowed for two sets of beams to be produced, each at one of two possible elevation angles. In contrast, a single on-axis array produces a set of azimuth scanned beams at a single elevation angle. In this geometry, each feed array is produced on a single planar circuit board, and so the feeds cannot be scanned to independent elevation angles.

In Reference 32, the feeds were of the tapered slot type and a demonstrator was constructed which generated two sets of eight azimuth scanned beams at 30 GHz from

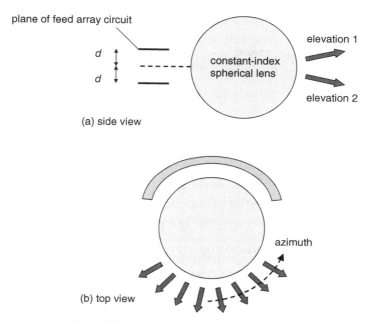

Figure 6.32. Off-axis array-fed spherical lens.

a 42-mm-diameter lens. The elevation scan angle was 15° and the azimuth beam spacing 16°, leading to a spatial coverage of 30° and 128°, respectively, for the set of 16 beams.

6.6 CROSS-POLARIZATION PROPERTIES OF SPHERICAL LENSES

A linearly and symmetrically illuminated aperture does not radiate cross-polarized energy. Of course, practical aperture antennas tend not to be illuminated with perfect symmetry, and so cross-polarized radiation is manifested as a consequence of geometric imperfections of the antenna aperture and/or the antenna feed. Compared to reflector antennas, lens antennas exhibit an advantage in this respect in that they lend themselves to symmetrical illumination by a primary feed at the lens focus yet without introducing the partial aperture blockage that would occur with the equivalent reflector optics. This property was mentioned in Chapter 5 in the context of low-sidelobe antennas for millimeter wave applications.

Spherical lenses retain this favorable symmetry and in theory exhibit zero cross-polarized radiation in the principal far-field planes. Nevertheless, in the nonprincipal planes, cross-polarization will occur. An example was studied in Reference 33, taking an electrically large aperture (60 cm at Ku band) as a test case. The SWE theory and computer code was again employed here to good effect. It was noted that an inherent property of the spherical vector wave modes is that cross-polarized terms vanish in the

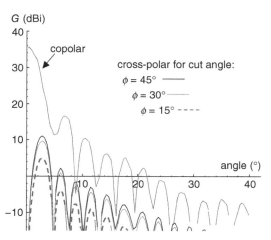

Figure 6.33. Cross-polarization in spherical lens, derived from SWE theory [33].

principal planes (*E*- and *H*-planes) and also on boresight for any plane. Trends are shown in Figure 6.33, where off-axis cross-polar radiation increases with the observation plane cut angle ϕ.

It should be noted that Figure 6.33 shows trends from theory, and a practical antenna tends always to exhibit cross-polar radiation levels higher than both theory predicts and oftentimes than is desirable in communications systems. Nevertheless, as long as care is taken to maintain symmetry of geometry of the lens and its feed, the spherical lens should score well in this respect.

REFERENCES

[1] C. A. Balanis, Antenna Theory, Analysis and Design, 3rd ed., John Wiley & Sons, Inc., Hoboken, NJ, 2005, pp. 300–304.

[2] R. J. Mailloux, Phased Array Antenna Handbook, 2nd ed., Artech House Antennas and Propagation Library, Norwood, MA, 2005.

[3] Q. Xu and J. Thornton, "Report on steerable antenna architectures and critical RF circuits performance," 2006. Deliverable D24 of the EU project "CAPANINA," Sixth Framework Programme, 2003–2006. FP6-IST-2003-506745, http://www.capanina.org

[4] R. K. Luneburg, Mathematical Theory of Optics, Brown University Press, Providence, RI, 1944, pp. 189–213.

[5] S. P. Morgan, "General Solution of the Luneberg Lens Problem," Journal of Applied Physics Vol. 29, No. 9, September 1958, pp. 1358–1368.

[6] H. Mosallaei and Y. Rahmat-Samii, "Non-uniform Luneburg and 2-Shell Lens Antennas: Radiation Characteristics and Design Optimization," IEEE Transactions on Antennas and Propagation Vol. 49, No. 1, January 2001, pp. 60–69.

[7] S. Cornbleet, "A Simple Spherical Lens with External Foci," Microwave Journal Vol. 8, No. 5, 1965, pp. 65–68.

[8] J. Thornton, "Scanning Ka-Band Vehicular Lens Antennas for Satellite and High Altitude Platform Communications," 11th European Wireless Conference, Nicosia, April 10–13, 2005.

[9] J. R. Sanford, "Scattering by Spherically Stratified Microwave Lens Antennas," IEEE Transactions on Antennas and Propagation Vol. 42, No. 5, May 1994, pp. 690–698.

[10] H. Schrank and J. Sanford, "A Luneburg Lens Update," IEEE Antennas and Propagation Magazine Vol. 37, No. 1, February 1995.

[11] M. Rayner, "Use of Luneburg Lens for Low Profile Applications," Datron/Transco Inc. Microwave Product Digest, December 1999.

[12] B. Schoenlinner, X. Wu, J. P. Ebling, G. V. Eleftheriades, and G. M. Rebeiz, "Wide-Scan Spherical-Lens Antennas for Automotive Radars," IEEE Transactions on Microwave Theory and Techniques Vol. 50, No. 9, September 2002, pp. 2166–2175.

[13] J. A. Stratton, Electromagnetic Theory, McGraw Hill, New York, 1942.

[14] A. C. Ludwig, "Spherical Wave Theory," in The Handbook of Antenna Design, Section 2.3, Vol. 1, A. W. Rudge, K. Milne, and A. D. Olver, eds., Peregrinus, London, 1982.

[15] A. C. Ludwig, "Near-Field Far-Field Transformations Using Spherical-Wave Expansions," IEEE Transactions on Antennas and Propagation Vol. AP-19, No. 2, March 1971, pp. 214–220.

[16] J. Thornton, "Scattering in Stratified Dielectric Lens Antenna with Covered Feed," International Journal of RF and Microwave Computer Aided Engineering Vol. 17, No. 6, November 2007, pp. 513–520.

[17] P. J. Wood, "Reflector Antenna Analysis and Design," Volume 7 of IEE Electromagnetic Waves Series, P. Peregrinus on behalf of the Institution of Electrical Engineers, 1986.

[18] N. Nikolic, "Scattering and Radiation in Spherical Structures," PhD Thesis, Department of Electronics of the Division of Information and Communication Systems, Macquarie University, Sydney, Australia, January 2008.

[19] N. Nikolic, J. S. Kot, and S. Vynogradov, "Analysis of Scanning and Multibeam Antennas Based on Spherical Lenses with Array Feeds," IEEE Vehicular Technology Conference (VTC), May 11–14, 2008.

[20] J. S. Gardner, "Approximate Expansion of a Narrow Gaussian Beam in Spherical Vector Wave Functions," IEEE Transactions on Antennas and Propagation Vol. 55, No. 11, November 2007, pp. 3172–3177.

[21] B. Fuchs, S. Palud, L. Le Coq, O. Lafond, M. Himdi, and S. Rondineau, "Scattering of Spherically and Hemispherically Stratified Lenses Fed by Any Real Source," IEEE Transactions on Antennas and Propagation Vol. 56, No. 2, Feb. 2008, pp. 450–460.

[22] CST MICROWAVE STUDIO®, User Manual Version 2009, September 2008, CST AG, Darmstadt, Germany, http://www.cst.com.

[23] D. Gray, J. Thornton, and R. Suzuki, "Assessment of Discretised Sochacki Lenses," Loughborough Antennas and Propagation Conference (LAPC), November 16–17, 2009.

[24] D. Gray, J. Thornton, H. Tsuji, and Y. Fujino, "Scalar Feeds for 8 Wavelength Diameter Homogeneous Lenses," IEEE International Symposium on Antennas and Propagation (AP-S 2009), Charleston, USA, June 1–5, 2009.

[25] N. Nikolic and A. R. Weily, "Realistic Source Modeling and Tolerance Analysis of a Luneburg Lens Antenna," IEEE Antennas and Propagation Society International Symposium (APSURSI), Toronto, July 11–17, 2010.

[26] J. Thornton, "Final Report of Multi-beam Scanning Antenna for Satellite Communications" under European Space Agency Contract no. 20836/07/NL/CB, ESTEC, The Netherlands, 2009.

[27] R. Donelson, M. O'Shea, and J. Kot, "Materials Development for the Luneburg Lens," International Square Kilometre Array Conference, Geraldton, Australia, July 27–August 2, 2003.

[28] J. Kot, R. Donelson, N. Nikolic, D. Hayman, M. O'Shea, and G. Peters, "A Spherical Lens for the SKA," Experimental Astronomy Vol. 17, No. 1–3, 2004, pp. 141–148.

[29] G. Befeki and G. W. Farnell, "A Homogenous Dielectric Sphere as a Microwave Lens," Canadian Journal of Physics Vol. 34, 1956, pp. 790–803.

[30] J. Thornton, "Versatility of Scanning Lens as Ground Antenna for HAP Communications: Low GHz to mm-wave," Meeting of COST-297, Nicosia, April 9, 2008.

[31] D. Gray, J. Thornton, and H. Tsuji, "Mechanically Steered Lens Antennas for 45 GHz High Data Rate Airliner-Ground Link," EHF-AEROCOMM/GLOBECOM 2008, New Orleans, USA, November 30–December 4, 2008.

[32] J. Zhang, B. Li, W. Wu, and X. Wu, "Wide-Scan Spherical Lens Antenna with Off-Axis Feed Arrays," IEEE APS Symposium on Antennas and Propagation, Charleston, USA, June 1–5, 2009.

[33] J. Thornton, "Sidelobe Analysis of Scanning Lens Antenna for Satellite Communications," Loughborough Antennas and Propagation Conference (LAPC), Loughborough, UK, March 17–18, 2008.

7

HEMISPHERICAL LENS-REFLECTOR SCANNING ANTENNAS

John Thornton

7.1 INTRODUCTION

This chapter reviews a range of activities and projects, carried out around the mid-2000s, concerning the development of hemispherical lens antennas for communications applications. Accordingly, the text and organization of this chapter tends to follow the chronology of that work.

These projects were applied in nature, and so much of the following content is concerned with both the design and the practicalities of fabricating the hardware encountered in this class of antenna. Once again, it is worth noting that the wider research programs, which lead to the *initiation* of this work, concerned communications from high-altitude platforms, and while Chapter 5 discusses lens antennas mostly in the context of HAP payloads and cellular networks, later research work in millimeter wave scanning antennas for vehicles identified the hemispherical lens reflector as a promising solution. Later, the research found its natural progression and practical application in satellite communications.

The scenario of initial interest was that of a communications terminal mounted on a moving vehicle, such as a high-speed train, providing a link to a HAP or satellite.

Modern Lens Antennas for Communications Engineering, First Edition. John Thornton and Kao-Cheng Huang.
© 2013 Institute of Electrical and Electronics Engineers. Published 2013 by John Wiley & Sons, Inc.

The two cases have much in common, but obvious differences include the very much reduced free-space loss and the increased variation of elevation angles, which would be encountered in the former case.

To list some of the characteristics of this scenario:

- Elevation angles could be between 90° and about 15°, and azimuth 0°–360°.
- Scan loss at low elevation is particularly disadvantageous since maximum aperture gain is needed here to counter the free-space and rain losses.
- The line-of-sight link could be subject to frequent blockages, and so a multibeam antenna could be useful to support link diversity.
- Low-profile antennas are advantageous on vehicles.

7.2 CANDIDATE SCANNING ANTENNA TECHNOLOGIES

The scope here is large and a very few candidates are considered that exhibit differing extremes of cost and complexity. Taking as a benchmark a medium- to high-gain antenna for 30 GHz and an antenna aperture diameter in the range of 150–400 mm, this respectively corresponds to a half-power beamwidth of about 5°-2° and a directivity of 32–40 dBi.* A mechanically steered aperture antenna would represent a pragmatic approach where it might be a reflector, horn-fed lens, or an array type. The reflector is conceptually one of the simplest solutions, and such an antenna would be expected to offer good radio frequency (RF) performance at a modest cost. The antenna would require housing within a transparent cover (radome) to protect it from adverse environmental conditions (wind loading, dust, snow). It would not offer multiple beams nor, typically, a low profile. Commercially available tracking antennas for Ku-band maritime applications most often use quite conventional parabolic reflectors and would fall within this class of antenna. Arrays, such as slotted waveguide arrays, are less often encountered in low-cost, commercial products and suffer from bandwidth and off-axis radiation problems.

Phased arrays were discussed in Chapter 6, where their performance and cost limitations were highlighted, particularly in the context of millimeter wave antennas. Nevertheless a number of programs in the 1990s and 2000s went some way toward developing digital beamforming (DBF) antennas for Ka band; for example, Dreher [1] reported in 2003 a 16-element module intended to eventually form the building block of a much larger aperture for aircraft–satellite links. The Japanese HAP program was at that time also developing DBF antennas for Ka band, and Miura and Suzuki [2] reported a 16-element prototype with three tracking beams.

Returning to the scenario of Figure 7.1 and the wide range of scan angles encountered, it is particularly apparent that a planar array conformal to the roof of a vehicle, that is, with the normal to the array plane pointing to the zenith (90°), will suffer quite

---

* For satellite communications, larger diameters would be needed for the data rates required by today's bidirectional "broadband" communications services.

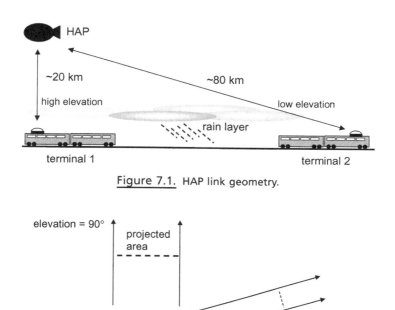

Figure 7.1. HAP link geometry.

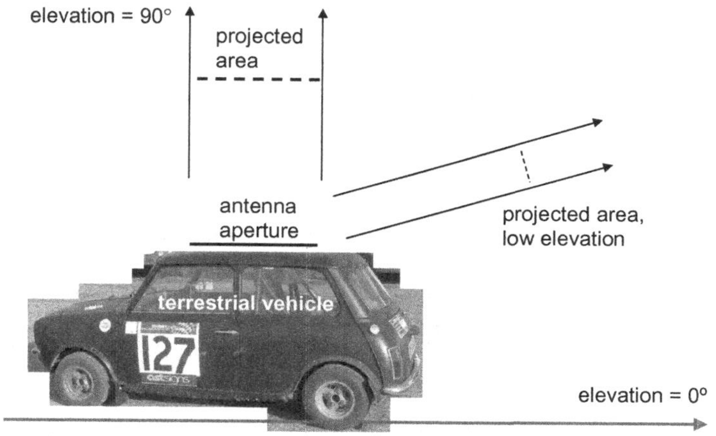

Figure 7.2. Reduction in projected aperture area with decreasing elevation angle.

badly from scanning loss for low elevation angles. This scanning loss is an inevitable consequence of the reduction in the array's projected area as elevation angle decreases (Fig. 7.2). As noted, for HAPs, communications link lengths are longest at low elevation, and so scanning loss here really cannot be tolerated [3].

Array scan loss can be estimated more rigorously by computing array directivity for scanned and nonscanned beams. Radiation patterns are presented in Figure 7.3, where the array is of a uniform rectangular layout and so exhibits the familiar first sidelobe at −13 dB. Introducing a properly calculated phase term at each element scans the beam—in this case, by 60° away from the array normal. It is evident here that the main lobe for the scanned beam is broader than for the nonscanned case. For both patterns, directivity has been calculated from the spatial numerical integration for total radiated power. The scan loss here amounts to 7 dB.

So, while a horizontally oriented planar array suffers scanning loss and mutual coupling causing "dead angles" as scan angle increases (elevation decreases), one might mitigate this effect by tilting the array to some extent away from the zenith and toward

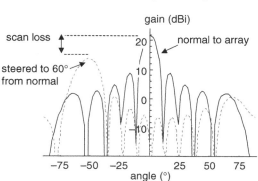

Figure 7.3. Array scan loss.

the horizon. This would allow the array to radiate more efficiently at whatever predetermined elevation angle was chosen. However, such an approach would then require a mechanical actuator to steer the antenna in azimuth since the array no longer illuminates a full hemisphere of space. A combined phased-array/mechanical hybrid scanning antenna is by no means an impractical proposition, but it is one that does tend to negate many of the advantages of beamforming antennas since a mechanical tracking stage would have to be retained.

7.3   SPHERICAL AND HEMISPHERICAL LENS ANTENNA

As discussed in detail in Chapter 6, the spherical symmetry of this class of lens allows it to employ multiple feeds for multiple beams and (to a first order) without introducing scanning loss. Beam steering may be achieved either by mechanically moving the feed or by switching between multiple feeds which are placed around the outside of the lens. A useful variant is a hemispherical lens used with a reflective ground plane. The purpose of the reflective ground plane is to produce an image of the hemisphere and so recovers the same effective aperture as a full spherical lens but occupying half the height (Fig. 7.4). This approach offers several advantages for a scanning, multibeam antenna placed on a vehicle:

- The lens does not have to be steered. The moving part (the feed) has much less mass than the lens.
- The physical height of the antenna and steering mechanism is approximately half that of a circular dish of equivalent gain.
- Multiple independent feeds, each illuminating a single and thus shared lens, can produce independent scanned beams.

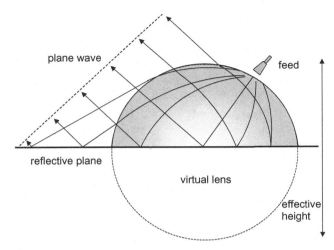

Figure 7.4. Effective aperture of hemispherical lens with ground plane.

- There is negligible scan loss; gain and beamwidth are constant irrespective of steered angle.
- Good power-handling capability.

7.4 HEMISPHERICAL LENS PROTOTYPE

The first of a family of polymer lenses for millimeter wave applications was reported by the author in Reference 4. This lens had a diameter of 160 mm and was machined from a single billet of polyethylene (PE), the same material as was used for the shaped low-sidelobe lens described in Chapter 5. This material is also known to have very low dielectric loss. Figure 7.5 shows this lens configured with a primary feed of waveguide horn type, detector (harmonically pumped mixer), and ground plane in an experimental setup in an anechoic chamber. The ground plane here is a square aluminum plate of side length 240 mm. At this stage of development, the purpose was to investigate scanning loss rather than to optimize antenna efficiency. Radio frequencies investigated were in the region of 28–30 GHz.

Initially, a pyramid waveguide horn was used for the primary feed. While this proved to be a quite poor feed for this application, the purpose of the experiment was primarily to investigate wide-scan properties using a limited ground plane area—the horn was used because it was available at the time. From this experiment, the measured half-power beamwidth was 7° from which the directivity was estimated at 29 dBi and represents an aperture efficiency of only 28%. This low efficiency was likely due to the combined effects of underillumination of the lens by the primary feed and defocusing caused by too large a distance between the horn phase center and the

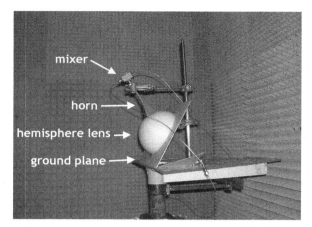

Figure 7.5. Constant-index hemisphere lens with feed at 65° elevation angle (photo courtesy of J. Thornton).

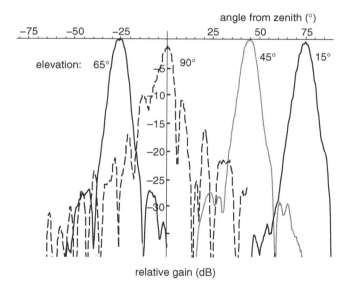

Figure 7.6. Experimental hemisphere lens with pyramid feed horn: measured radiation patterns for four elevation angles.

lens edge. Ohmic loss in the PE material was thought to be a very much smaller source of loss.

Figure 7.6 shows measured radiation patterns for four cases of elevation angle and showing minimal scan loss (0.7 dB compared to peak gain) for a 15° elevation angle and a 1.8-dB degradation for 90°, that is, the zenith beam (dashed line).

The beam degradation at zenith is due to the primary feed partially blocking the antenna aperture, leading to gain reduction and increased sidelobe levels. This aperture blockage effect would be reduced either by the use of a larger lens or a more compact primary feed. It is quite likely that a polyrod feed of one of the types described in Chapter 3 could be used to good effect here because this type of feed, being dielectric, presents a smaller cross section than an air-filled horn. In any case, the zenith effect is not problematic for many practical scenarios: For HAP communications, the zenith directed link is also the shortest (lowest free-space loss), and for satellite communications, elevation angles are seldom near the zenith unless the ground station operates in equatorial regions. In this latter case, the ground plane might be preinclined so that the feed never needs to be directly overhead. The experimental arrangement of Figure 7.5 is clearly not a very practical one for deployment on a vehicle—the feed components are cumbersome and are far from ideal for both packaging and RF purposes. Nevertheless, it represented a useful investigative tool and served as a first step toward more advanced stepped-index polymer lenses and a deeper understanding of them.

An alternative type of primary feed based on a "scalar" or "choked" circular waveguide feed was next used with the above constant-index lens. This feed was designed according to the recipe of Olver et al. [5], where four concentric grooves of quarter-wave depth were machined into the waveguide flange. This yields a broader and more symmetric primary feed pattern (Fig. 7.7) than that of the pyramid horn, has a phase center very close to the aperture, and so illuminates the lens more effectively.

The choked, scalar feed was subjected to some quite extensive analysis, also using the solver FEKO, which agreed extremely closely with the far-field patterns generated using CST MICROWAVE STUDIO (CST MWS). Then, a series of measurements were carried out at various frequencies and for both principal planes—one comparison between measurement and theory is shown in Figure 7.8.

This detailed analysis of the feed was valuable in providing confidence in the various solver tools available, the measurement system, and allows the use of the feed as a calibration standard when later measuring the gain of the feed–lens combination. Using this feed with the 160-mm PE lens, the measured gain was around 30 dBi at 28 GHz, representing an aperture efficiency of approximately 40% (Fig. 7.9).

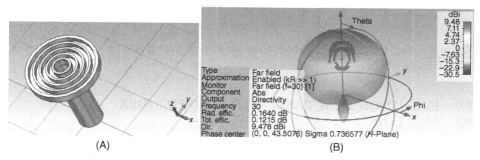

(A)   (B)

Figure 7.7. Scalar feed (9.5 dBi) for 30 GHz (A) geometry and (B) simulated radiation pattern from CST MWS.

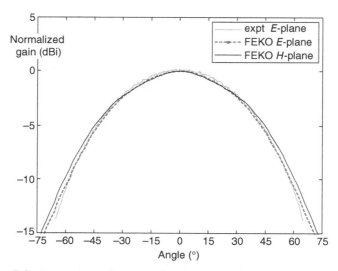

Figure 7.8. Comparison of measured and FEKO simulated radiation pattern.

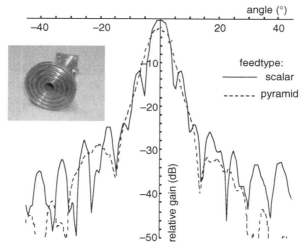

Figure 7.9. Measured lens antenna radiation patterns comparing feed types (from Reference 4; inset: image of scalar feed).

7.5 EVOLUTION OF A TWO-LAYER STEPPED-INDEX POLYMER LENS

Building on the first experimental steps discussed above for the constant-index lens, and the spherical wave expansion (SWE) technique discussed in Chapter 6, research continued in 2004–2005 to explore what practical lens antennas of the stepped-index

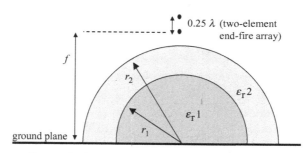

Figure 7.10. Two-shell hemisphere lens geometry.

TABLE 7.1. Dielectric Materials Considered for Two-Shell Hemisphere Lens

Material	ε_r	Loss Tangent (tan δ)
Polyurethane foam	1.2	$<10^{-4}$
Polyethylene	2.28	8×10^{-4}
Rexolite	2.53	7×10^{-4}
Fused silica	3.8	7.5×10^{-4}

type might be feasible to construct in the laboratory. This work was inspired in part by the reported two-layer spherical lens geometry of Mosallaei and Rahmat-Samii [6], where the dielectric constant was considered as a variable parameter that could be fine-tuned to two decimal places for purposes of optimization. In contrast, a more pragmatic approach was sought [4] where a short list of available, low-loss dielectric materials was considered in various combinations in a two-layer stepped-index spherical lens (Fig. 7.10 and Table 7.1).

In Section 6.4, the properties of constant-index spherical lenses were discussed and where it was noted that efficiency can be good up to moderate lens electrical diameter, beyond which efficiency declines. In this regime, increasing the number of dielectric layers pays dividends. As a reference point against which to quantify the efficiency improvement of a two-layer lens, the efficiency of a constant-index lens should serve as the benchmark and where the focal distance should have the same value in both cases. Fixing the focal distance (f in Fig. 7.10) allows a like-for-like comparison for lens antennas occupying a given space.

Figure 7.11 shows the sensitivity to f for eight wavelength radius constant-index spherical lenses of different materials. (The diffraction-limited directivity of a uniform $8\,\lambda$ aperture is 34.7 dBi.) The trend that lower ε_r materials exhibit a greater focal distance is expected [7]. However, during this investigation, it was noticed that the "paraxial focus" from Schoenlinner et al. [7],

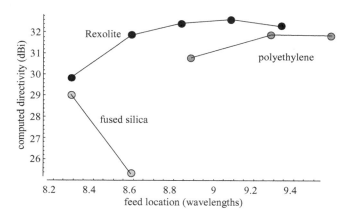

Figure 7.11. Spherical lens computed directivity D versus feed location for three dielectric materials (after Thornton [4]).

TABLE 7.2. Directivity and Focal Distances of Spherical Lenses

	Maximum D (dBi)	Feed Location (λ) for Maximum D	D for Fixed 8.2 λ Feed Location
Polyethylene	31.9	9.2	31.3
Rexolite	32.7	9.0	32.0
Fused silica	29.1	8.2	29.1

$$f = \frac{\sqrt{\varepsilon_r}}{2\left(\sqrt{\varepsilon_r} - 1\right)},$$

leads to a greater value of f than the value derived from the SWE method when used to search for maximum directivity D. (A detailed discussion of the relationship between dielectric constant and focal length is also found in Reference 8.)

Higher values of focal distance f lead to an increase in the required height for the antenna installation. In Reference 4, a minimum feed–lens separation of 0.2 λ was used. This places the feed close to the lens edge but allows up to 0.2 λ displacement between the feed's physical outer aperture and its phase center. Such a gap should be allowed for experimentation and tuning purposes when the antenna is tested in practice—it is unwise to design for the condition f = lens radius (zero gap)!

The maximum directivity D for the three cases in Figure 7.11 is shown in Table 7.2 along with the feed position at which this was achieved. The last column then shows the case when the feed is fixed at 8.2 λ so that the comparison is like for like in terms of the antenna's occupied space. In deriving this result, the lens diameter was reduced by a necessary amount so that the focal position was maintained.

Table 7.2 provides a benchmark against which a stepped-index lens may be compared. Using a similar SWE code as was used for constant-index lenses, various two-shell lenses with the same fixed focal position of 8.2 λ were investigated. The outer radius r_2 was fixed at 8 λ and r_1 was allowed to vary. Materials were chosen from the short list of Table 7.1, but always with $\varepsilon_{r1} > \varepsilon_{r2}$.

In summary, this work found:

- There is no advantage to adding an outer layer to a fused silica core.
- There is minimal advantage in using the low-constant foam as an outer layer. The layer would need to be very thin, of the order of one wavelength, and hence difficult to fabricate.
- The best results are obtained using a Rexolite inner core of radius 4.2 λ and a PE outer layer. This yields 33.46 dBi, which represents 76% aperture efficiency (albeit with an idealized feed model).

This analysis, using available materials in (more or less) trial-and-error combinations, and the SWE computer code, found that a two-layer lens of 8 λ outer diameter would offer in theory a 1.46-dB gain improvement over a single-layer (homogeneous) lens. (A caveat to these results is a reminder that the feed model is simplistic: Should a model be used which allows the feed to be tailored to a particular lens diameter and dielectric constant, slightly different results could be expected, albeit at the cost of a considerable increase in computational complexity.)

For larger-diameter lenses, the advantage of the two-layer lens becomes greater. In another example, for a focal length of 11.5 λ, the directivity of a single Rexolite spherical lens is maximized at 34.4 dBi for a radius of 10.35 λ, while the two-shell Rexolite/PE combination offers 36.1 dBi when the lens outer radius is 11.1 λ, being a 1.7-dB improvement. These dimensions translate to a 236-mm diameter at 28 GHz, a frequency of interest for terrestrial millimeter wave communications. This design was taken forward into an experimental stage and reported in Reference 9, the dimensions being tractable for machining and available material billets. Some fabrication issues were encountered apparently stemming from shrinkage or other distortion of the PE polymer billet after it was machined into the concave hemisphere, leaving a 300-μm gap between the edge of the hemisphere's flat face and the ground plane. The gap was reduced by abrasion. The Rexolite did not exhibit a noticeable deformation, and although this part did not contain a concave bowl, which probably contributed to the deformation of the PE, Rexolite would appear to be a more stable material that is better behaved when machined.

The fabricated dielectric layers are shown in Figure 7.12. (They are not particularly photogenic.)

The measured gain of the two-layer polymer lens, illuminated by the scalar waveguide feed, was 35.1 dBi $\pm$ 0.4 dB, which represents an aperture efficiency in the region 68%. This efficiency is comparable with a reflector antenna and suggests that dielectric loss is not a very significant factor. The measured radiation pattern is shown compared

Figure 7.12. HDPE and Rexolite layers for 236-mm stepped-index lens.

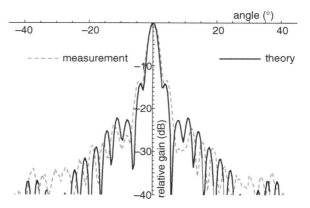

Figure 7.13. Two-layer lens: *H*-plane patterns at 28 GHz and 2.7-m measurement distance (after Thornton [9], © IET).

to the theory in Figure 7.13. Here, it should be noted that these are not far-field patterns. Owing to the limited length of the anechoic chamber, the measurement distance (between antenna under test and reference antenna) of 2.7 m was somewhat less than the far-field condition would dictate. This effect was accounted for in the numerical computation where this value is used in the evaluation of the spherical Bessel functions instead of their far-field asymptotic values.

Thornton [9] also proposed a refinement to the SWE theory for a hemispherical lens compared to a full sphere. It had been observed that sidelobe patterns in some cases agreed poorly with theory and also changed according to the feed inclination angle with respect to the ground plane. This should not be surprising. However, it should also be accepted that a deficiency in the SWE theory, which has been used so far in this text, is an oversimplistic feed model and a source of error should be expected

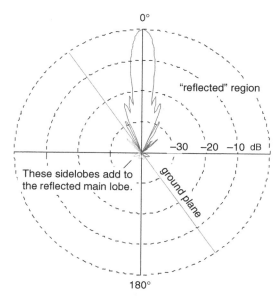

Figure 7.14. Visible and reflected regions arising from reflector.

here. Nevertheless, it was proposed that a better description for the scattering properties of the hemisphere with the ground plane could be obtained by first considering the pattern of a spherical lens as being cut into two halves. Of these, one contains the main lobe, which is reflected by the ground plane, and one contains only sidelobes. Then, by superposition of the two regions into one-half of space, by vector addition, the radiation pattern for the hemispherical lens could be obtained. These regions are illustrated in the polar plot of Figure 7.14, where the orientation of the ground plane with respect to the main lobe selects that portion of the sidelobe pattern, which adds into the main lobe region.

The net result of this method led to a negligible effect on the main lobe and the theoretical antenna efficiency but was found to lead to a better agreement with measured sidelobe patterns. This effect is well illustrated in Figure 7.15, which shows E-plane patterns and compares spherical lens theory (SWE), hemisphere theory (modified SWE with reflected region as described above), and measurement. The finiteness of the ground plane was not addressed at this stage, although the near-field distance effect was included in the SWE terms, and so a direct comparison between simulation and measurement can be made for the near field. In Figure 7.15, we see that for this particular plane and feed orientation, the hemisphere theory better predicts the measured sidelobe pattern than does the more simplistic spherical lens theory. In contrast, H-plane patterns were found to be very little affected because the additive sidelobe level was much lower than in the E-plane case.

Theoretical far-field patterns for the two-layer lens according to the method of [9] are shown in Figure 7.16.

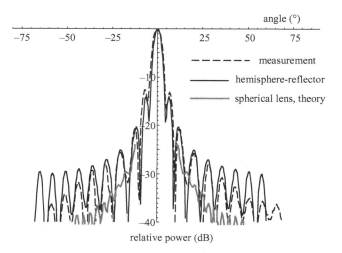

Figure 7.15. Constant-index hemisphere lens-reflector radiation patterns: measurement and theory (after Thornton [9], © IET).

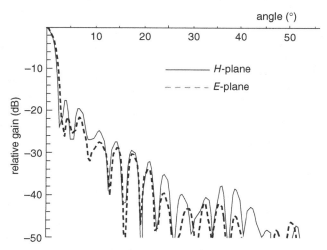

Figure 7.16. Far-field radiation patterns from SWE, two-layer hemispherical lens (after Thornton [9], © IET).

7.6   A HEMISPHERICAL LENS-REFLECTOR ANTENNA FOR SATELLITE COMMUNICATIONS

The success of the two-layer polymer lens promoted some interest from the satellite communications community and this area became a logical direction for exploiting what so far had been research spun out of the HAP activities touched upon in Chapter

5. Antennas for satcoms on the move (land mobile) were highly topical in the mid-2000s and continue to be at the time of writing, particularly bearing in mind the rapid growth in available bandwidth at Ka band. The relatively low profile of a hemisphere lens (more correctly, "lens reflector" since the reflective ground plane is an intrinsic component in the aperture) makes it quite attractive for vehicular applications. In 2007, the European Space Agency (ESA) supported a study in scaling the stepped-index lens antenna to demonstrate its feasibility for using a Ku-band satcom on the move. The main objectives were

- fabricating a larger aperture lens comparable to very small aperture terminal (VSAT) or ultrasmall aperture terminal (USAT) and
- demonstrating mechanical beam scanning for two independent feeds.

These two objectives could loosely also be described as a work package in RF development and another in the electromechanical scanning system. Performance at Ku band was expected to be good because the electrical size of the lens would be comparable to that of the preceding experimental lens at 30 GHz. A secondary objective was to explore performance at Ka band (20 and 30 GHz).

The intention was to demonstrate a Ka-band beam and a Ku-band beam, each being independently steerable. This offers the service operator greater flexibility in terms of which satellites may be used. For example, a service offered to coach or train passengers might include satellite TV in a receive-only mode, but should a two-way internet service be required as well, a single-beam antenna would tie the provider to a satellite at the same orbital location. However, a second steerable beam allows these additional services to be streamed via some other satellite. A similar situation is found in maritime applications and offshore industries where satellite is the only viable communications technology. Broadcast, internet, and communications services might be procured from a wide variety of orbital locations, leading to a proliferation of antenna terminals. Sometimes, duplicate terminals are used because vessel superstructures limit the field of view. In such cases, multibeam antennas can greatly reduce installation costs.

A field trial or live demonstration of the proposed antenna on a moving vehicle was, however, beyond the scope of this development.

7.6.1 Requirements

The frequency requirements for the dual-band lens antenna were those shown below in Table 7.3. These bands were chosen to be representative of those typically used for land mobile VSAT/USAT services. A "minimum" requirement means one that can be used by a specific satellite service and frequency plan, while a "target" just means that as wide a bandwidth as possible should be addressed so that the antenna could be used with (almost) any frequency plan. In practice, in VSAT antenna feed chains, bandwidth constraints tend to come about from items such as orthomode transducers, filters, polarizers, and amplifiers, and the development of a full-feed chain was beyond the scope of this work, where the emphasis and novelty fell upon the lens aperture. Accordingly, the requirements were used to guide the range of frequencies over which

TABLE 7.3. Frequency Requirements for Dual-Band, Dual-Beam Antenna

	Ku-Band Frequency (GHz)	Ka-Band Frequency (GHz)
Uplink minimum target	14.0–14.25	29.5–30.0
	12.7–14.5	
Downlink minimum target	10.7–12.75	19.7–20.2
		18.3–20.2

it was desirable to measure the properties of the lens. Later, nominal feeds were used, which slightly limited the range of frequencies used in practice.

The aperture size chosen was 61 cm. This is toward the smaller size of aperture that would be used at Ku band but could be quite representative of a Ka-band on-the-move terminal.

The range of elevation scan angles was derived from the minimum satellite elevation of 20° and maximum of 55°, and the maximum measured train roll angle of 7°, leading to a scan requirement of 13°–62°. These figures are applicable to relatively high latitudes and would cover Europe.

The angular scan rates were the following:

- Maximum rate of roll, pitch, and yaw: 5°/s
- Maximum azimuth acceleration: 4°/s^2
- Maximum elevation acceleration: 1°/s^2

7.6.2 Lens Analysis

Again using the SWE computer code, the following theoretical lens efficiencies were derived for a two-layer lens with an outer diameter of 610 mm. The location of the feed was also adjusted to search for the optimum location. In practice, a compromise feed position would have to be fixed. Table 7.4 shows efficiency for the three main frequency bands of interest, and here it is apparent that the efficiency falls away quite sharply at higher frequencies.

Table 7.5 shows several frequency points within Ku band. Here, the lens should work very well, as expected from earlier experience with the 236-mm scale model.

A predicted radiation pattern is shown in Figure 7.17 and compared with an industry standard (Eutelsat Standard M), which places a limit on off-axis radiation. Here, an advanced version of the SWE feed model has been included, which seeks to better approximate a realistic feed. The greater aperture taper function (narrower primary feed beamwidth) is manifested, as expected, in slightly suppressed sidelobes.

7.6.3 Three-Layer Lens Geometry

In an attempt to improve efficiency at the higher frequencies of interest, particularly at 30 GHz, a three-layer lens structure was investigated from theory. This entailed a lengthy series of computer simulations. As electrical size increases (the 610 mm equates

TABLE 7.4. 610-mm Diameter, Two-Layer Lens Characteristics from SWE Simulation

Frequency (GHz)	Layer Radii		Focal Distance		Directivity (dBi)	Aperture Efficiency (%)
	r_1 (λ)	r_2 (λ)	r_feed (λ)	(mm)		
12	5.86	12.2	12.7	320	36.8	82
20	9.60	20.33	21.5	322	40.2	63
30	14.4	30.5	32.5	325	41.6	40

TABLE 7.5. Lens Theoretical Performance at Ku Band

Frequency (GHz)	r_2 (λ)	Focal Position		Directivity (dBi)	Aperture Efficiency (%)
		r_feed (λ)	(mm)		
10.7	10.9	11.33	319	36.0	84
11.7	11.9	12.6	323	36.6	83
12.7	12.9	13.7	322	37.2	80
13.25	13.5	14.2	322	37.5	79
13.75	14.0	14.7	321	37.8	78
14.5	14.7	15.5	320	38.2	77

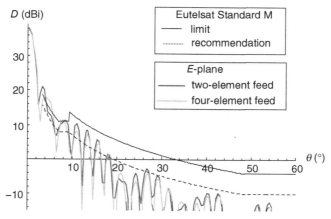

Figure 7.17. Copolar patterns at 14.5 GHz.

to 61-wavelength diameter), the number of modes in the SWE must be increased, and this adds to simulation time. We are also presented with far more variables—choice of dielectric constant for the three layers and the first two-layer radii.

The geometry is shown in Figure 7.18.

A two-feed model was again used, this requiring approximately 40 minutes of processor time to derive aperture efficiency for each set of input parameters. The aperture efficiency was taken as the sole figure of merit for the optimizations.

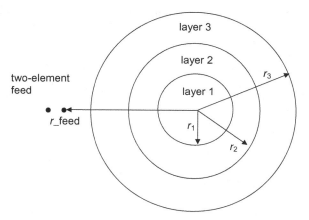

Figure 7.18. Three-layer model.

TABLE 7.6. Dielectric Constant Sets for Three-Layer Geometry Radii Optimization Runs in SWE, 30 GHz

Data Set	ε_{r1}	ε_{r2}	ε_{r3}
1	2.6	2.4	2.0
2	2.6	2.2	2.0
3	2.5	2.4	2.1
4	2.4	2.4	2.1
5	2.6	2.4	2.1
6	2.5	2.2	2.1
7	2.5	2.2	2.0
8	2.7	2.6	2.4
9	2.5	2.4	2.3

TABLE 7.7. 610-mm Three-Layer Lens, Theoretical Efficiency

Frequency (GHz)	D (dBi)	Aperture Efficiency (%)
12	35.8	65
20	40.0	61
30	43.5	63

The input data sets are summarized in Table 7.6; that is, random choices of dielectric constant were fixed, then the radii were treated as variables in an optimization routine. It is not desirable to treat the dielectric constant as a parameter that can be fine-tuned unless one has access to exotic synthetic materials. In any case, the values tabulated do not in every case correspond to identifiable materials.

The best geometry is data set 3, which predicts 63% aperture efficiency. An analysis of this structure at the other frequencies of interest yields the data shown in Table 7.7.

The results in Table 7.7 are to some extent unexpected because the three-layer structure, albeit derived from an attempt at optimizing for 30 GHz, gives inferior performance at 12 GHz compared to the two-layer design. The performance at 20 GHz for the two- and three-layer designs is similar. The main advantage of the three-layer design presented here is a useful increase in efficiency at 30 GHz. However, these results are the products of a limited investigation and, as has been noted, difficulties obtaining materials curtailed the search for a three-layer design.

7.6.4 Lens Fabrication and Performance

In parallel with the above three-layer analysis, an investigation into low-cost air–polymer foam materials was being carried out in an attempt to synthesize these intermediate dielectric constants in Table 7.6. Some of those results were discussed in Chapter 6, but ultimately, synthetic materials were not used. Attempts to purchase syntactic foam were also unsuccessful. This led to the continued use of solid polymer dielectric materials and the established two-layer design using PE and Rexolite.

The outer PE layer proved the most difficult to fabricate. Here it was found necessary to join together two 300-mm-thick billets to form a pseudobillet, which was then machined into the concave outer hemisphere. The inner Rexolite layer was machined according to the description in Section 6.4.1. The inner and outer layers are shown in Figure 7.19, here nearing completion. The heads of six PE bolts can be seen in the face of the PE layer, and the six holes in the Rexolite layer have yet to be plugged.

Later, the flatness of the completed PE layer had to be improved, again using an abrasive flat bed in a similar manner as had been found necessary for the 236-mm scale model.

The completed 61-cm-diameter hemisphere lens was placed on a flat aluminum reflector (ground plane), and using a conventional, domestic television receive only

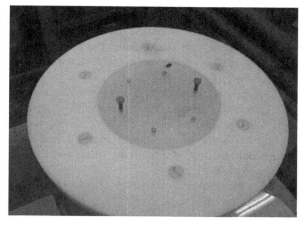

Figure 7.19. The 61-cm two-layer polymer lens under construction.

Figure 7.20. Receiving satellite TV with the lens (photo courtesy of J. Thornton).

TABLE 7.8. Lens and Dish Dimensions

	Lens	Dish
Dimensions (m)	Diameter = 0.61	Minor axis = 0.59
		Major axis = 0.64
Area (m²)	0.2922	0.2965
Diffraction limit on D (dBi) at 12.0 GHz	37.69	37.75

(TVRO) satellite receiver low-noise block (LNB), the main lobe pointed at the broadcast satellite. This represented the first really convincing demonstration that the new two-layer lens was successful. The lens with feed was compared to the USAT parabolic dish antenna as seen in Figure 7.20 and here, the half-height comparison is most striking. This was a practical demonstration that the lens could in fact be used to receive the broadcast signal and that the lens could match the electrical performance of the usual offset parabolic dish antenna. There was no disadvantage or loss of performance from halving the antenna height.

This demonstration really amounted to a straightforward comparison with the dish rather than a more rigorous or calibrated measurement. The aperture areas are very similar but not exactly so: The dish is slightly elliptical, having the dimensions shown in Table 7.8. As such, it has a very slightly larger aperture than the lens and a correspondingly higher maximum theoretical directivity, although any difference is too fine to be detected in measurements.

Within the tolerance of measurement accuracy of power at intermediate frequency (IF), the dish and lens had the same gain from 10.7 to 12.5 GHz. In this comparison, the same primary feed was used so as to eliminate differences in LNB output power. The midband gain was estimated at 35.6 dBi.

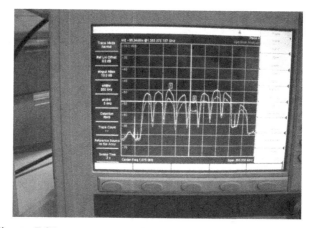

Figure 7.21. Transponder spectrum: Astra 2 broadcast satellite.

A commercially available seven-layer quasi-Luneburg lens of conventional construction, with a diameter of 61 cm, was also used for comparison. It is just visible on the left side of Figure 7.20. A typical satellite transponder spectrum is shown in Figure 7.21, where the lower trace is that of the Luneburg lens and the upper trace that of the two-layer polymer lens. These traces indicate the commercial item performed about 4 dB worse, depending on frequency; however, it is estimated that in this case, about 1.5 dB is due to the different LNB output powers. Other measurement uncertainties include loss in the windows of the building, which could not be removed, and the different ground planes used: Although these had very similar areas, they may have had different surface flatness. A reasonable estimate of the performance difference between the commercial lens and the two-layer lens is the 2-dB advantage to the latter for 10.7–12.5 GHz.

Later measurements with the completed scanning antenna, that is, after integration of the lens with the ground plane and mechanical scanning system, were carried out using an improvised outdoor range. These proved difficult owing to the antenna mass. In one suite of measurements, the main lobe pattern was captured by sweeping the feed through an arc calculated to form a great circle in antenna spherical axes. From this, the main lobe half-power beamwidth could be derived. Also, measurements with antenna feeds of known gain provided calibration data. Errors, however, came about from pointing loss and less than optimum feed focal point. These errors were worse at Ka band, where the sensitivity to feed position is more acute. The measurement results shown in Table 7.9 represent a best estimate having aggregated various measurement data.

7.6.5 Mechanical Tracking System

In parallel with the lens analysis and development, a mechanical tracking system was designed and a prototype built. This activity, also supported by the ESA development,

TABLE 7.9. Hemispherical Polymer Lens Reflector: Measured Gain

Frequency (GHz)	Gain (dBi)
10.7	35.1
11.5	35.5
12.5	36.0
20.0	38.5
30.0	40.0

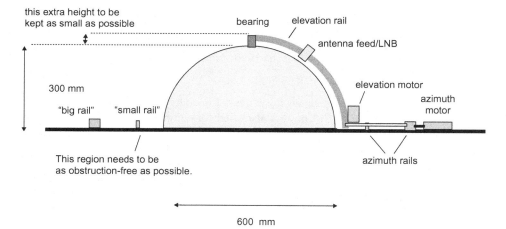

Figure 7.22. Scanning lens antenna: an early conceptual sketch.

took place over a roughly 18-month period starting 2007. Its starting point was a "clean sheet," but some driving requirements were established at the outset. Central to these were a desire to minimize antenna height, which would preclude placing any components underneath the reflector plane and the avoidance of scattering by components in the proximity of the main lobe. Accordingly, motorized carriages would move around the circumference of the lens whereby the motors and primary feeds moved together and so leaving the opposite side of the lens clear of obstruction. The rail to which these carriages were mounted would inevitably contribute to scattering, and so this was placed some distance from the lens edge.

Taking the starting point of Figure 7.22, a suite of mechanical components was designed in proper detail. These entailed computer numerical control (CNC) machined elevation supports, or "arms," which would carry both the primary feeds and timing belts by which the feed was actuated. At the base of each arm, on an azimuth carriage, brushless DC servo motors would provide azimuth and elevation actuation. The motors, by Maxon, were used with in-line reduction gearboxes and integrated position encoders.

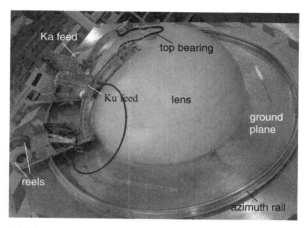

Figure 7.23. Dual-beam scanning hemisphere lens-reflector antenna.

A single circumferential azimuth rail supported the two carriages, their elevation arms, and feeds. This rail was fabricated from eight similar CNC machined aluminum components. The rail would also house four cables (three RF coaxial cables, one multicore cable for DC power and control). These cables would wrap under slight tension around the outer perimeter of the rail and be carried by sprung reels to spool the cables as the carriages move in azimuth. Coaxial rotary joints were housed within the rails to route IF circuits, while a multiway slip ring handled power and can control bus signals. The mechanical system was developed in parallel with the polymer lens and, to some extent, in isolation from it since a dummy hemispherical lens was used to support various build iterations. Figure 7.23 shows an almost complete build of the scanning system with the polymer lens.

Azimuth actuation was via rack and pinion. The toothed rack was housed on the inside of the azimuth rail. Magnetic sensors were added on the azimuth and elevation rails so as to provide positional feedback. These were used to characterize the accuracy and reliability of the motor system and, if necessary, could be retained within a closed loop control system. One finding of this work was that the accuracy and repeatability of the motor and mechanical system were such that the additional position feedback provided by the magnetic sensors would not necessarily be needed.

7.6.5.1 Feeds. The development of purpose-made feeds for the lens reflector was considered beyond the scope of this project. Quite conventional feeds were considered adequate to demonstrate the usability of the scanning lens antenna. To this end, the TVRO LNB was all that was needed to show that the antenna worked well at Ku band as we have seen above in the reception of satellite TV. This feed is also pictured in Figure 7.24, mounted to one of the two elevation arms. An improved geometry would be one where the LNB electronic circuit is angled at 90° with respect to the feed waveguide axis, rather than in line as pictured. This would reduce headroom at high scan

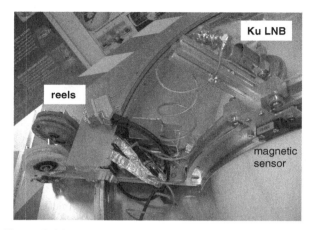

Figure 7.24. Ku-band TVRO feed on scanning lens reflector.

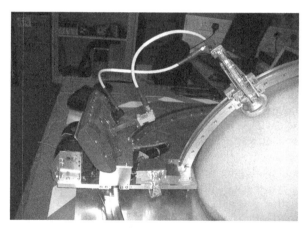

Figure 7.25. Modified Ka-band transceiver and OMT.

angles. Such a "folded" feed is entirely practicable but was not pursued during this phase of development.

It remained to demonstrate the use of a two-way beam at Ka band. A low-cost Ka-band terminal intended for domestic satellite services in North America was modified slightly by adding flexible cables between the ortho-mode transducer (OMT) and transceiver transmit and receive ports. This is shown in Figure 7.25. Here, two disadvantages were apparent: As noted above, a folded waveguide geometry would reduce headroom, and the cables will contribute to RF losses. In fact, these were very high quality cables but would still degrade the noise figure at the intended 20-GHz receive frequency by around 1 dB. This could be ameliorated by moving the low noise amplifier (LNA) circuit from the transceiver and into the OMT. Again, this practical modi-

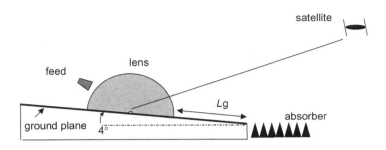

Figure 7.26. Measurement of ground plane truncation loss.

fication was considered somewhat outside the scope of this lens antenna development program.

In Figure 7.25, the azimuth motor can just be seen as the black cylinder in the center of the picture: It is aligned along a diagonal with respect to the azimuth carriage and drives a pinion, via a 90° gear, which engages the azimuth rack. The elevation motor is out of view on the underside of the azimuth carriage.

7.6.6   Ground Plane Effects

To avoid truncation of the effective aperture at a given elevation angle e, the required minimum rectangular ground plane extension length L_{min} is given by

$$L_{min} = \frac{r_{lens}}{\sin(e)} - r_{lens},$$

where r_{lens} is the lens radius. The loss due to lesser values was investigated at a single elevation angle by measuring the received power in the satellite broadcast signal as illustrated in Figure 7.26.

This is a quite straightforward measurement technique: It is necessary only to vary the ground plane extension L_g by varying the position of the lens. The 61-cm polymer lens described above was used. The ground plane was preinclined by 4°, leading to an effective elevation angle of 27°. The use of an absorber, as illustrated, helps reduce edge-scattering effects.

The measured loss data for two polarization cases E_v and E_h (vertical and horizontal, respectively) are plotted in Figure 7.27. Also shown (dashed line) is the area reduction that occurs from the truncation of the projected circular aperture. The measured loss is somewhat less than the area reduction because the part of the aperture that is truncated is weakly illuminated due to the tapered aperture distribution.

A more detailed analysis of ground plane effects followed [10, 11] since this is an important property of the hemisphere lens reflector. It was proposed in References 10 and 11 that the ground plane truncation loss could be estimated by considering the near field of the aperture and by simply "deleting" that section of the aperture which was truncated, and then performing a transform to the far field and computing directivity.

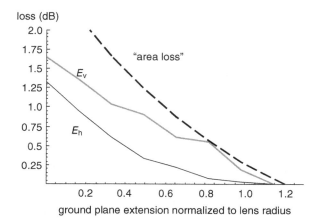

Figure 7.27. Measured ground plane truncation loss.

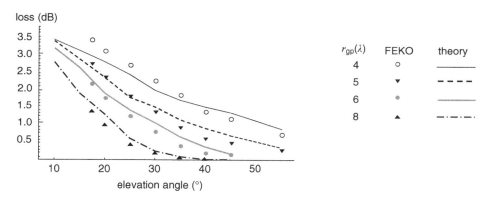

Figure 7.28. Ground plane truncation loss versus elevation angle (after Thornton et al. [11], © IET).

Of course, it is first necessary to have derived the near-field aperture distribution, and here, the SWE technique, which has already been described, can readily be used—it is necessary to use the near-field terms for the spherical Bessel and Hankel functions in place of the far-field versions. This is computationally time-consuming, but the far-field transform that follows is very much less so. This means that having derived the near fields in some arbitrary plane close to the aperture, this data set can be manipulated by striking out regions that are truncated by arbitrary shapes, be they circular or square ground planes or any other structure. Edge scattering is not accounted for by this approximation. A full description is found in Reference 11.

A parallel approach was taken using the solver FEKO and results compared with the SWE-transform method described above. Some results are shown in Figure 7.28 for an eight-wavelength-diameter constant-index lens.

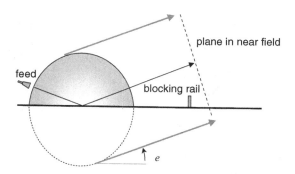

Figure 7.29. Lens aperture blockage by rail.

7.6.7   Aperture Blockage in Scanning Lens Reflector

Ground plane truncation is not the sole enemy of the hemisphere lens reflector antenna. While this occurs at low elevation, the orientation of the feed can begin to cause blockage of the aperture at high elevation angles (i.e., elevation is here referred to as the angle between the antenna boresight and the ground plane). The antenna works best for intermediate scan angles where neither ground plane truncation nor feed blockage occurs. This can be a wide range of angles and easily sufficient to meet the needs of the satellite communications antenna being considered. A further problem, however, arose from the azimuth rail (Fig. 7.23), which has already been alluded to, this partially blocking the aperture. An interesting way of avoiding this completely would be to fit the rail on the underside of the ground plane, but because this would add to the antenna height, that solution had already been rejected at the outset. The SWE and far-field transform computation was entirely practicable for estimating the blockage effect: It was necessary to derive the blocked region of the aperture by projecting the rail dimensions against it for a given elevation angle, as illustrated in Figure 7.29.

This problem was too large to attempt to use in a solver such as FEKO or CST MWS. Also, measurements with the full size antenna were quite inconvenient and, in any case, some insight was sought into what would be gained by reducing the rail size or altering its position. In contrast, measurements with a scale model were much easier, and so these were carried out to augment the theoretical findings. Figure 7.30 shows one such arrangement where the azimuth rail is represented using a strip of printed circuit board. The position and height of this component could be changed quite easily so as to gather a range of data points. The frequencies used were in the region of 28–31 GHz, which would scale to 10.8–12.0 GHz for the full size antenna, and polarizations both parallel and normal to the ground plane were investigated.

Some data are shown in Table 7.10, where the 14-mm rail height is scaled from 36 mm for the full size component. At an 18° elevation, the scan loss was worse and could amount to 1.5 dB.

As expected, the loss reduced as the rail height was reduced or its distance from the lens edge increased. Loss was seen to vary quite weakly with frequency but more

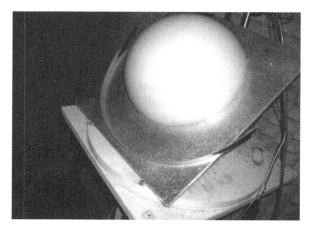

Figure 7.30. Scale model of lens reflector with azimuth rail (photo courtesy of J. Thornton).

TABLE 7.10. Loss from Rail Blockage for 24° Elevation and at 31 GHz

Rail Height (mm)	Loss (dB) Parallel		Loss (dB) Normal	
	Theory	Measurement	Theory	Measurement
14	0.79	0.9	0.96	1.2
10	0.58	0.7	0.62	0.6

strongly with polarization. A rail height of 13 mm (full scale) or 5 mm (high-frequency model) saw the loss at 18° elevation not exceed 0.7 dB, and this could be further reduced if the distance from the lens edge could be increased from the value of 175 mm used in the full size antenna. The height of the rail had been driven largely by the need to route four cables about its circumference, stacked in two pairs (one cable pair for each feed). Tidying the cable routing strategy, for example, by adopting optical fiber for the IF signals (all the IF bandwidth could be aggregated onto a single fiber), would thus have the very useful benefit of directly improving scan performance for this type of multibeam lens antenna.

7.7   A LOW-INDEX LENS REFLECTOR FOR AIRCRAFT COMMUNICATIONS (CONTRIBUTION BY D. GRAY)

A constant-index hemisphere lens reflector was proposed for aircraft-to-ground communications by Gray et al. [12, 13] at Japan's National Institute of Information and Communications Technology (NICT).

The proposed application was a 100-Mbps duplex Q-band (43.5–47.0 GHz) radio link between jet airliners and fixed ground stations (Fig. 7.31). Compared to a satellite

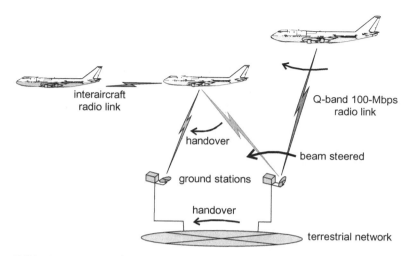

Figure 7.31. System sketch of proposed Japanese domestic jet airliner connectivity system.

communications terminal, the aperture may be somewhat smaller owing to the much reduced link distance. In an effort to reduce mass, temperature cycle effects, and to avoid stringent manufacturing tolerances, a low-index, low-density, dielectric material was investigated. The target antenna gain was 26 dBic across and 7° 3-dB beamwidth across 43.5–47.0 GHz [14]. The required across-track scan range (perpendicular to the direction of flight) was ±48°, while along-track was ±70°. The latter scan requirement is acknowledged as being beyond the capability of a planar phased-array antenna, dictating either a multiple feed switched topology such as that in Reference 15 or a more compact and lower-cost fully mechanically steered antenna. As this was a civilian project and application, the latter option was chosen. Further, to avoid the expense of rotary joints, the feed was to remain fixed, while the hemispherical lens and flat reflecting plate would be rotated by a small (150-mm diameter) commercially available two-axis gimbal (Fig. 7.32). This arrangement intrinsically solves the severe along-track requirement as the antenna is fully rotationally symmetric in that plane. The challenge is to design an antenna with sufficient gain in the across-track plane across the full 7% bandwidth. The maximum allowable width of the lens antenna was taken to be the diameter of the gimbal; 150 mm is 20 λ at 45 GHz.

The two-shell design of Section 7.5, scaled to a radius of 3.65 wavelengths, would meet this requirement and serve as a benchmark. A constant-index lens with equivalent gain was sought since this would reduce mass and manufacturing complexity. A material with $\varepsilon_r = 1.7$ was chosen since a low-density foam of this specification was offered commercially: ECCOSTOCK LoK foam by Emerson and Cumming Microwave Products. The specific gravity of the material is 0.54, which compares well with the polymers used in an earlier work where the figure is close to 1. The only disadvantage of this material was its relatively high loss tangent (tan $\delta = 0.004$), which was expected to cause approximately 1-dB loss in the lens. (The analytical formula for loss is taken

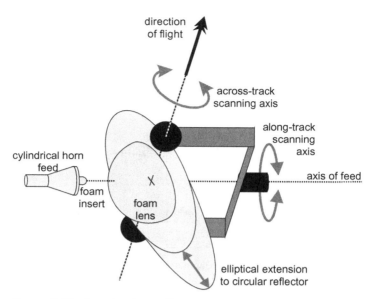

Figure 7.32. Scanning lens reflector geometry, from Gray et al. [12].

from and is as Eq. 4.22 already presented in the section on millimeter wave antennas in Chapter 4.)

Thus, the material was judged useful as a low overhead cost surrogate for an initial experimental investigation of the antenna geometry and a low-loss foam would be developed later if the antenna entered serial production. The geometry is shown in Figure 7.32.

A lens radius of four free-space wavelengths (λ) was chosen. This was conveniently analyzed using the commercially available simulator FEKO running on a modest desktop personal computer. The relatively small size of the antenna, compared to those discussed previously, combined with the efficiency of representing the lens with the method of moments surface equivalence principle and solving by FEKO's Multilevel Fast Multipole Method solver gave simulation times of about 5 hours per frequency step and required 9 GB of RAM.

A number of factors dictated that the initial prototype antenna should be a scaled model to be tested at X-band. Chief among these were that all NICT anechoic chambers, except the one at Kashima Space Center, were in constant use. The Kashima Space Center chamber had been used for satellite work in the past and was well equipped with X-band horns and support equipment. Further, the lens, feeds, and support jigs for a scaled model at X-band could be manufactured internally for the cost of materials only, while a Q-band antenna would have to be subcontracted out at a considerable expense to an optical instrument maker. Thus, the initial prototype antennas were designed for 12 GHz, and measurements would be made across 11.5–12.5 GHz to confirm the full bandwidth characteristics; the scaling factor was 3.75. The 4 λ radius

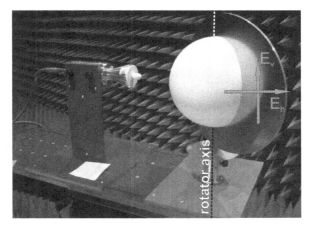

Figure 7.33. LoK foam lens with "insert" primary feed (photo courtesy of D. Gray/NICT).

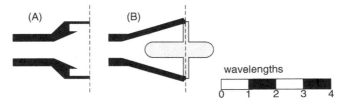

Figure 7.34. Cross sections of feed antennas trialed for the LoK lens: (A) Wolf feed [18] and (B) insert feed [12]; the dashed lines mark the reference plane.

lens consequently had a diameter of 200 mm, and both it and the jigs were easily transported and handled (Fig. 7.33).

In the initial FEKO simulations, only a circular ground plane was used, of extension 2 λ beyond the lens edge, and gave poor gain performance at wider scan angles [12]. This first version of the antenna fell short of the target gain by about 1 dB. Various innovative measures were used, which improved the antenna efficiency. These included

- examination of alternative feeds including a dielectric rod feed,
- addition of an elliptical ground plane extension, and
- addition of a cylindrical disk section (step) at the lens base.

Five candidate feeds were initially considered [16]. Three were well-known scalar parabolic reflector feeds. The other two were a short backfire antenna and a novel circularly symmetric dielectric rod design based on a prior linearly polarized design [17]. The underlying concept of this latter design was that a low dielectric constant lens has a diffuse cigar-shaped focus, so logically, the most appropriate feed would have a matching near field. A 15-mm-diameter cigar of LoK foam 62.5 mm long suspended in the mouth of a small conical horn was found to be the optimal design in FEKO. Holding the cigar in the mouth of the horn with a 4.8-mm (a quarter of a guided wavelength)-thick sheet of LoK foam did not affect performance (Fig. 7.34). This novel

TABLE 7.11. Measured Feed Antenna Characteristics at the Scaled Center Frequency (12 GHz)

Feed	Feed Diameter (λ)	Separation from Lens Surface (λ)	3-dB Beamwidth (°)		10-dB Beamwidth (°)		Received Gain (dBi)
			E	H	E	H	
Insert	2.4	1.9	43	41	91	75	12.2
Wolf	2.2	1.5	35	33	67	66	13.9

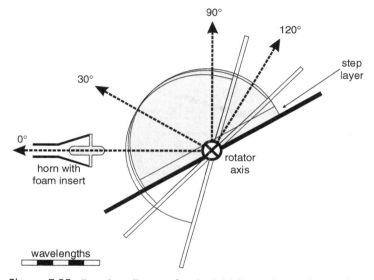

Figure 7.35. Top view diagram for the initial experimental campaign.

feed and one of the scalar feeds proved superior as feeds to the LoK foam lens and are considered here in some detail.

The scalar feed was a small dual-mode horn [18] and had a slightly smaller diameter than the conical horn with a LoK foam insert (Fig. 7.34 and Table 7.11). Despite the smaller size, the scalar horn had narrower beamwidths and higher gain.

With the hemispherical LoK lens and backing circular ground plane added to the circularly polarized insert feed FEKO model, a series of simulations were run to investigate the across-track scan performance. The across-track scanning angle was defined as the main lobe direction from the axis of the feed, with 0° defined as the main lobe pointed directly back at the feed (Fig. 7.35). Rotating the lens and ground plane at 45° would point the main lobe to 90°, for example.

The first set of FEKO simulations for the LoK foam lens with circular ground plane extending 2 λ beyond the lens edge nearly satisfied specifications (Fig. 7.36). An ellipti-

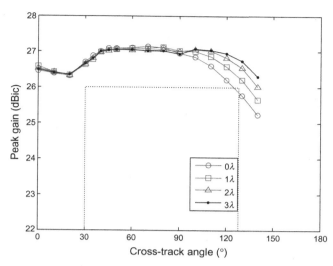

Figure 7.36. FEKO peak gain variation with cross-track scan angle of LoK lens fed by insert feed for different elliptical extensions beyond the 2 λ circular ground plane, 12.0 GHz, 0 λ was circular.

cal distortion to one-half of the circular ground plane was added in 1 λ steps and was found to increase the peak gain beyond an across-track angle of 90°. An elliptical extension of 3 λ was judged to be adequate, and further extensions were not investigated [12, 13]. A ground plane with a 3 λ (75-mm) extension was used for all experiments (Fig. 7.33).

Linearly polarized radiation pattern measurements were undertaken in the 11.5-m anechoic chamber at Kashima Space Center. Despite the application calling for circularly polarized antennas, measuring the linearly polarized components enabled quick fault finding and analysis. As expected, measured received gain levels were approximately 1 dB below the level expected. A theoretical loss of 0.9 dB was calculated for a 4 λ radius lens of dielectric constant 1.7 with tan $\delta = 0.004$ from the homogeneous lens dielectric loss equation in Reference 19. This value was added to all received gain measurements to give the gain values presented; hence, the results of Figure 7.37 are referred to as "adjusted" here. The soundness of this approach is supported by the high level of agreement between the measured/adjusted gain values and patterns and those derived from FEKO simulations in Figure 7.38, for example. (Ultimately, a lower-loss material would be needed to replace the LoK surrogate.)

A small number of across-track radiation patterns from the LoK lens fed by the Wolf feed at the center frequency are presented here to show the reasonable level of agreement found between measurement and the FEKO simulations (Figs. 7.38 and 7.39). As expected from the structural asymmetry of the antenna in the across-track plane, the radiation patterns were asymmetric on either side of the main lobe. The position of radiation pattern nulls and level of sidelobes was in good agreement between

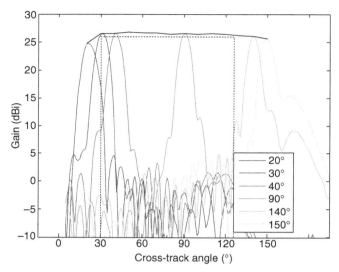

Figure 7.37. Measured and adjusted radiation patterns of the LoK lens fed by the Wolf feed with peak gains, 12.0 GHz, E_v polarization.

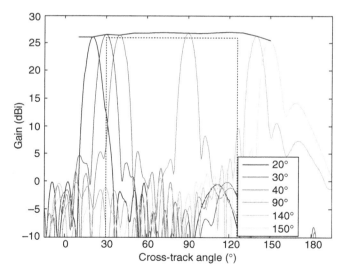

Figure 7.38. FEKO radiation patterns of the LoK lens fed by the Wolf feed with peak gains, 12.0 GHz, E_v polarization.

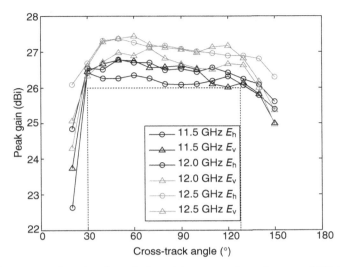

Figure 7.39. Measured peak gain variation with cross-track angle across 8.3% bandwidth of LoK lens fed by Wolf feed; dashed line marks gain-scan requirements.

FEKO and measurement. Of particular interest is the high spillover sidelobe close to 180° for a scan angle of 150°; that sidelobe is relatively high, contains a high level of power leading to noticeable gain loss, and could be mitigated by increasing the elliptical ground plane extension. The only major differences were for a scan angle of 20°, where the measured radiation pattern had lower gain (1 dB lower), broader 3-dB beamwidth (0.5°), and different sidelobes (at 40°) (Figs. 7.37–7.42). The likely cause of this difference was that the thickness of the ground plane was not accounted for in the FEKO simulations, where the ground plane was assumed to be infinitely thin.

For both linear polarizations, the LoK lens with Wolf feed surpassed the gain requirement for the entire 8.3% bandwidth studied (Fig. 7.39). As a generalization, across all across-track angles, the gain was higher at the upper end of the band studied. This fits with the expectation that a larger aperture antenna will produce a higher gain. Both the measured and FEKO 3-dB beamwidths were 1°–2° wider than the specification of 7° (Figs. 7.41 and 7.42); this would not preclude the use of this antenna for the stated application. Also, as a generalization, for the majority of across-track angles, the 3-dB beamwidth was narrower at the upper end of the band where the antenna was electrically larger.

Due to time and budget constraints, a double-nested gimbal jig could not be built, so radiation pattern measurements in the E_v-plane of Figure 7.33 could not be undertaken. In contrast, it was exceptionally straightforward to extract radiation patterns from FEKO in the plane perpendicular to the across-track scanning plane (Fig. 7.43). Although it is perhaps a little misleading without some explanation, the along-track plane radiation patterns are presented in the same format as the across-track patterns in Figure 7.38. (These two-dimensional radiation patterns should be thought of as

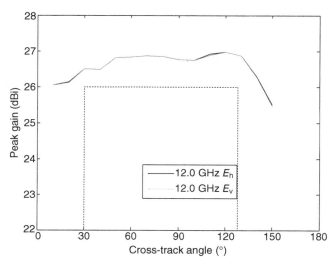

Figure 7.40. FEKO peak gain variation with cross-track angle across at 12.0 GHz of LoK lens fed by Wolf feed; dashed line marks gain-scan requirements.

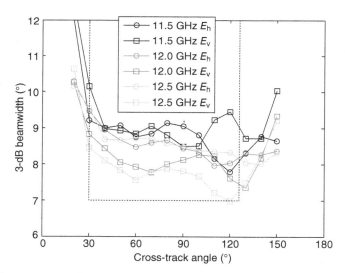

Figure 7.41. Measured 3-dB beamwidth variation with cross-track angle across 8.3% bandwidth of the LoK lens fed by the Wolf feed; dashed line marks gain-scan requirements.

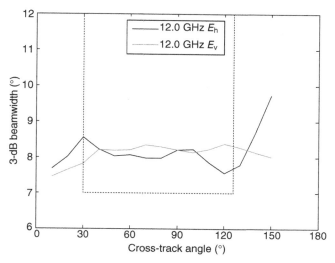

Figure 7.42. FEKO 3-dB beamwidth variation with cross-track angle at 12 GHz of the LoK lens fed by the Wolf feed; dashed line marks gain-scan requirements.

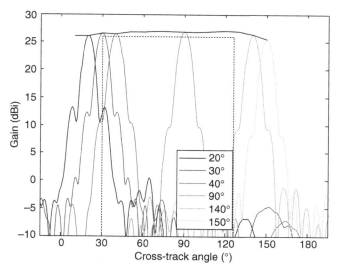

Figure 7.43. FEKO along-track plane radiation patterns of the LoK lens with Wolf feed, 12.0 GHz, E_v polarization.

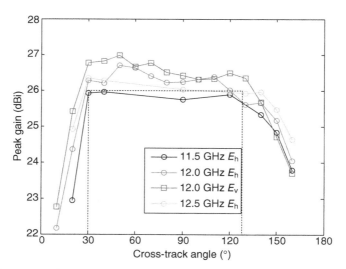

Figure 7.44. Measured peak gain variation with cross-track angle across 8.3% bandwidth of LoK lens fed by insert feed; dashed line marks gain-scan requirements.

oriented out of the plane of the page and should not be lying in it.) All of the radiation patterns in the along-track plane were found to be symmetrical, as expected from antenna structural symmetry, and the sidelobe level was generally around 17 dB below peak as would be expected of a circular aperture with uniform illumination; the lens was neither under- or overilluminated in that plane.

Changing to the small conical horn with the suspended LoK foam insert gave a slightly better gain at the center frequency of 12 GHz (Fig. 7.44). Similarly, the 3-dB beamwidths at 12 GHz were a little narrower than with the Wolf feed (Fig. 7.45). There was little difference between feeding the LoK foam lens with either of the feeds at the center frequency other than the insert feed having a longer focal distance (Table 7.11). The significant difference between the two feeds is in performance away from the center frequency. At both 11.5 and 12.5 GHz, the insert feed LoK foam lens failed to meet gain specification for the wider scan angles. As the lens showed adequate performance with the other feed, these results show that the insert feed is too narrowband for this application. It is speculated that increasing the lens radius to 4.5 λ and reoptimizing the dielectric cigar-shaped rod may result in a specification compliant design. Alternatively, the lens shape could be distorted or optimized to better suit the insert feed. As FEKO is a full-wave simulator, the shape of the lens was not constrained to spheres as it was with the SWE method; any shape can be simulated.

In a similar vein to the circularly polarized FEKO simulations investigating the effect of elliptical ground plane extensions, thin cylindrical steps or layers were added between the lens and the ground plane (Fig. 7.35). These simulations were done with the 6 λ radius circular ground plane only. Although introducing the steps did not improve the gain at wider across-track scan angles, the steps did improve the shallower

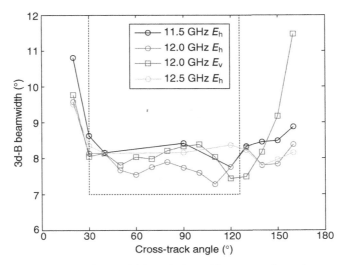

Figure 7.45. Measured 3-dB beamwidth variation with cross-track angle across 8.3% bandwidth of LoK lens fed by insert feed; dashed line marks gain-scan requirements.

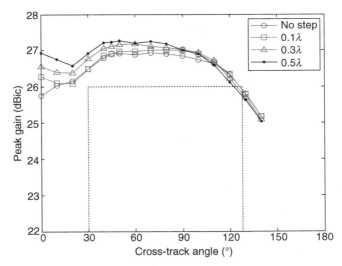

Figure 7.46. Effect of the step height on scan performance of LoK lens fed by the insert feed, from FEKO with 6 λ circular ground plane only.

angle gain (Fig. 7.46). A 9.5 mm (0.4 λ) layer of LoK foam was available, and the across-track plane radiation patterns were measured (Fig. 7.47). For scan angles of 30°–60°, the gain across the entire 8.3% bandwidth was increased to 27 dBi or better. However, the wider angle gain was little changed compared to the simple hemispherical LoK foam lens. It is speculated that these initial lens shape distortion–optimization

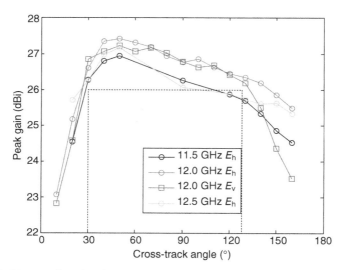

Figure 7.47. Measured peak gain variation with cross-track angle across 8.3% bandwidth of LoK lens with step fed by insert feed; dashed line marks gain-scan requirements.

effects can be improved upon in the future through the use of fully automated multivariable optimization with FEKO and that the optimal shape will be symmetric only about the across-track plane.

The measured performance of a 4 λ radius homogeneous lens with a dielectric constant of 1.7 was found to meet specification with one feed and to be close to meeting specification with the other. An elliptical ground plane extension was shown to improve gain performance at a wide-angle scan, while a simple lens distortion was shown to improve gain performance at shallow angles. With a diameter of 12 λ, the ground plane of the Q-band version of this lens antenna would easily fit into a small 150-mm (20 λ) off-the-shelf gimbal.

Due to reduced Japanese Public Service budgets, the original plan to demonstrate the Q-band ground to jet airliner communications systems was scaled back considerably. As a result, this lens antenna work was terminated at the end of the 2008/2009 financial year. During the 2009/2010 financial year, a small-scale demonstration was conducted over Haleiwa, North Shore, O'ahu, Hawaii using a small privately owned propeller-driven aerobatic planes flying at an 8000-m altitude [20]. The onboard antenna used was a phased array in the across-track plane while been mechanically steered in the along-track plane [14]. The antenna was a novel design composed of 16 silver-plated low temperature co-fired ceramic (LTCC) dielectric waveguide subarrays, each having 16 open waveguide radiators. Each radiator had an individual perturbation for narrowband circular polarization, which restricted operation to far less than the 7% bandwidth of Q-band. For the flight demonstration, the antennas were installed in an external belly pod [20] and were set into the aircraft belly as originally proposed for the jet airliners [21]. A data rate of 100 Mbps between the aircraft and a ground station was shown during the demonstration flights. Further, some common Internet applications such as web browsing, e-mail, and a video call by personal computer webcam operated normally in the aircraft cockpit [20].

REFERENCES

[1] A. Dreher, "Digital-Beamforming Antenna for Broadband Communications in Ka-Band," 6th International Symposium on Wireless Personal Multimedia Communications (WPMC 2003), Japan, October 19–22, 2003.

[2] R. Miura and M. Suzuki, "Preliminary Flight Test Program on Telecom and Broadcasting Using High Altitude Platform Stations," Wireless Personal Communications Vol. 24, No. 2, 2003, pp. 341–361.

[3] J. Thornton, "Antenna Technologies for Communications Services from Stratospheric Platforms," 4th International Airship Convention and Exhibition, Cambridge, UK, July 28–31, 2002.

[4] J. Thornton, "Scanning Ka-Band Vehicular Lens Antennas for Satellite and High Altitude Platform Communications," 11th European Wireless Conference, Nicosia, April 10–13, 2005.

[5] A. D. Olver, P. J. B. Clarricoats, A. A. Kishk, and L. Shafai, Microwave Horns and Feeds, IEEE Press, New York, 1994.

[6] H. Mosallaei and Y. Rahmat-Samii, "Non-uniform Luneburg and 2-Shell Lens Antennas: Radiation Characteristics and Design Optimization," IEEE Transactions on Antennas and Propagation Vol. 49, No. 1, January 2001, pp. 60–69.

[7] B. Schoenlinner, X. Wu, J. P. Ebling, G. V. Eleftheriades, and G. M. Rebeiz, "Wide-Scan Spherical-Lens Antennas for Automotive Radars," IEEE Transactions on Microwave Theory and Techniques Vol. 50, No. 9, September 2002, pp. 2166–2175.

[8] S. S. Vinogradov, P. D. Smith, J. S. Kot, and N. Nikolic, "Radar Cross-Section Studies of Spherical Lens Reflectors," Progress in Electromagnetics Research Vol. 72, 2007, pp. 325–337.

[9] J. Thornton, "Wide-Scanning Multi-layer Hemisphere Lens Antenna for Ka Band," IEE Proceedings—Microwaves, Antennas & Propagation Vol. 153, No. 6, December 2006, pp. 573–578.

[10] J. Thornton, A. D. White, and D. Gray, "Multi-beam Lens-Reflector for Satellite Communications: Construction Issues and Ground Plane Effects," European Conference on Antennas and Propagation (EUCAP09), Berlin, pp 1377–1380, March 23, 2009.

[11] J. Thornton, S. Gregson, and D. Gray, "Aperture Blockage and Truncation in Scanning Lens-Reflector Antennas," IET Microwaves, Antennas & Propagation Vol. 4, No. 7, July 2010, pp. 828–836.

[12] D. Gray, J. Thornton, and H. Tsuji, "Mechanically Steered Lens Antennas for 45 GHz High Data Rate Airliner-Ground Link," EHF-AEROCOMM/GLOBECOM 2008, New Orleans, USA, November 30–December 4, 2008.

[13] D. Gray, J. Thornton, and H. Tsuji, "Mechanically Steered mm-Wave Aeronautical Lens Reflector Antennas—Development of 45 GHz Homogeneous Lens Reflector Antenna for Civilian Airliners," Meeting of IEICE Technical Committee on Antennas and Propagation, National Defense Academy, Japan, September 11, 2008.

[14] M. Hieda, H. Tsuji, K. Matsukawa, M. Miyazaki, and Y. Aramaki, "Broadband Radio Communication System for Aircraft Using Millimeter-Wave Band," Mitsubishi Electric Corporation Technical Journal Vol. 84, No. 11, November 2010, pp. 621–624.

[15] W. Williams, "Advanced Lightweight Electronically Steered Antennas for Responsive Space Payloads," AIAA-LA Section 1st Responsive Space Conference, Redondo Beach, CA, April 1–3, 2003, pp 1–10.

[16] D. Gray, J. Thornton, H. Tsuji, and Y. Fujino, "Scalar Feeds for 8 Wavelength Diameter Homogeneous Lenses," IEEE Antennas and Propagation Symposium Digest, North Charleston, June 2009, paper 429.6.

[17] F. Averty, A. Louzir, J. F. Pintos, P. Chambelin, C. Person, G. Landrac, and J. P. Coupez, "Cost-Effective Antenna for LEO-Satellites Communication System Using a Homogeneous Lens," IEEE AP Symposium, 2004, pp. 671–674.

[18] H. Wolf and E. Sommer, "An Advanced Compact Radiator Element for Multifeed Antennas," 18th European Microwave Conference, 1988, October 1988, pp 506–511.

[19] Y. T. Lo, S. W. Lee, Antenna Handbook: Antenna Applications, Vol. 3, Springer, New York, 1988, pp. 17–19, Chapter 17.

[20] H. Tsuji, M. Suzuki, and T. Morisaki, "Millimeter Broadband Wireless System Aims to Accommodate Inflight Wireless Communications," The 13th International Symposium on Wireless Personal Multimedia Communications (WPMC 2010), Recife, Brazil, 2010.

[21] D. Gray, "Lens Antenna Apparatus," Japanese Patent JP2010034754, granted February 2, 2010.

ABOUT THE AUTHORS

John Thornton graduated in physics and worked for some time in the aerospace industry, which he then left to return to academia. He completed a Master's degree in microwave physics in 1995, worked for a time at the Rutherford Appleton Laboratory, and later took up a research post at the University of Oxford. He then became a Research Fellow at the University of York, a post held for some 10 years—much of the content of this book is taken from that period—and completed his PhD at the Open University. He is a member of IEEE. In 2010, he joined MDA Space and Robotics Ltd and leads their U.K. antennas program.

Kao-Cheng Huang, PhD, is Vice President at the Dharma Academy in Taiwan and an IEEE Senior Member. Dr. Huang formerly worked as the MSc program leader at the University of Greenwich, UK, and as senior research engineer at Sony Technology Centre, Germany. He received his PhD degree from the University of Oxford, UK, and is the author of two books related to millimeter wave technology. He holds several world and international patents in the area of millimeter wave communications.

Modern Lens Antennas for Communications Engineering, First Edition. John Thornton and Kao-Cheng Huang.
© 2013 Institute of Electrical and Electronics Engineers. Published 2013 by John Wiley & Sons, Inc.

INDEX

Modern Lens Antennas for Communications Engineering, First Edition. John Thornton and Kao-Cheng Huang.
© 2013 Institute of Electrical and Electronics Engineers. Published 2013 by John Wiley & Sons, Inc.